EARTHQUAKE: A NATURAL DISASTER

EARTHQUAKE: A NATURAL DISASTER

ASHUTOSH GAUTAM

A P H PUBLISHING CORPORATION
4435-36/7, ANSARI ROAD, DARYA GANJ
NEW DELHI-110 002

Published by
S.B. Nangia
A P H Publishing Corporation
7, Ansari Road, Daryaganj
New Delhi 110002
Ph.: 23274050
E-mail : aphbooks@gmail.com

2024

Rs. 3195/-

Printed at
Balaji Offset
Navin Shahdara, Delhi 110032

Dedicated to
the peoples died
in 20th Oct. 1991
and 30th Sept. 1993
Earthquakes

Preface

Geologically, 55% area of India is highly seismic i.e. prone to earthquakes. Seismologists have explained the origin of earthquakes on the basis of plate tectonics. Earthquakes of 20th Oct. 1991 and 30th Sept. 1993 struck the region of Uttarkashi district of Uttar Pradesh and Latur and Osmanabad districts of Maharashtra in a davastating manner. These regions suffered mostly due to the unscientific and unplanned development.

Keeping in mind that it will not occur in future during the occurrence of such or bigger quakes we are presenting this volume. This is a collection of articles have been written after the earthquakes of 20th Oct. 1991 and 30th Sept. 1993. All the articles deal with one basic theme, and the title of the book essentially describes it.

The contribution highlight the important issues and problems created by earthquakes, like — origin of earthquakes, casualties and damage, developments after earthquakes, dams in relation to earthquakes, impact of earthquakes on ecology, environment, agriculture, prediction of earthquakes and protection measures.

We have taken every precaution to ensure that information presented in this volume is accurate and uptodate. However we cannot accept any responsibility for errors and omissions.

ASHUTOSH GAUTAM

Acknowledgements

I wish to thank all the people who helped in bringing out this volume precisely I desired. Special thanks are due to 'The Hindustan Times Ltd.' for giving permission to publish various articles and photographs appeared in Hindustan Times.

It is my pleasure to acknowledge the kind help, received from Prof. Ram Kumar Sharma, Mrs. Vinod Bala Sharma, Dr. N. Singh, Alok, Aditya and Anamika. Thanks are also due to Mr. Vinod Joshi, Mr. C.P. Juyal, Mr. O.P. Semwal, Mr. O.P. Gusain and Mr. Krishna Kumar Upreti. My wife, Anjali, has helped me greatly during the preparation of this volume.

Thanks are also due to M/s. Ashish Publishing House, Delhi, for bringing out the volume with an elegant get-up.

ASHUTOSH GAUTAM

CONTENTS

①

Earthquake An Introduction

Amongst all natural calamities earthquake is the most devastating and dangerous. Beginning of earthquake is closely related to the origin of earthquake and formation of earth surface. Origin of earth took place million years ago on detachment of one part from the Sun, which cooled down slowly-slowly and solidification of upper part took place. The upper part (Surface) is near about 180 Km thick. On this upper surface todays world is existing. The earth crust could be divided into six big (few small) plates which are floating over a semi liquid portion. Due to this fact these plates are in moving condition. When these plates colloid together, a pressure originates which in-turn shake the earth surface, known as earthquake.

EARTHQUAKE AND INDIA

Fifty five per cent area of India is earthquake prone, which had faced many earthquake of varying intensities in the recent past, but still the people of these regions are unknown to this fact. According to the seismicity India could be divided in to three major parts i.e. the region having high sensitivity to earthquakes, the region having moderate sensitivity to earthquakes and the region having low sensitivity to earthquakes (Fig. 1.1). As per geologists/seismologists the Himalayan part is more sensitive to earthquakes because of the fact that it is the youngest mountain range only 1-2.5 crore years old, and the youngest mountain ranges are quite prone to the appearance of earthquakes.

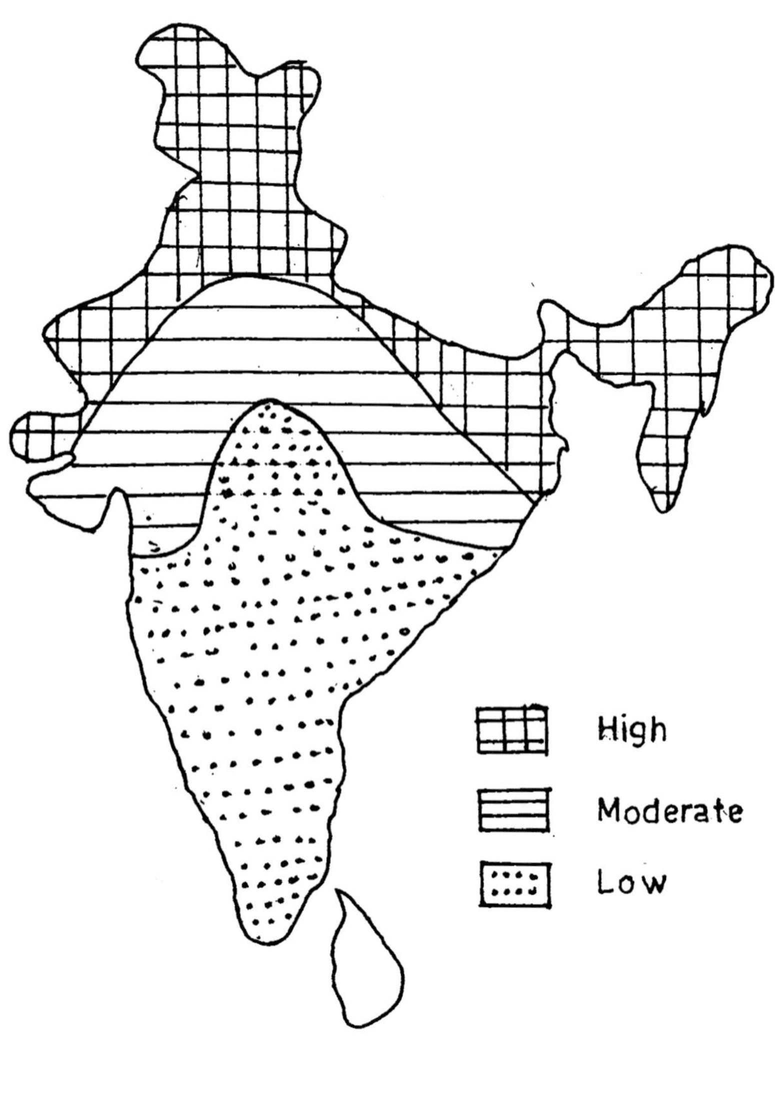

Fig. 1.1 Sensitivity to Earthquakes

India's Himalayan part which includes Srinagar, Leha, Laddakh, Kumaon, Garhwal, Kohima, Imphal, Gauhati, Bondila, Agartala etc. and the planes of Ganga and Indus are quite prone to earthquakes.

HISTORY OF EARTHQUAKES

Before the appearance of recent earthquake of 30th September 1993, India has faced a quite good number of earthquakes. Garhwal faced an earthquake on 20th October 1991, it was measured 6.1 on Richter scale, remained for 50 seconds and resulted in the death of more than 1500 peoples. On 20th August 1988 Bihar faced an earthquake of 8.1 on Richter scale, the death toll was near about 1000. On 10th December 1987 Maharashtra faced an earthquake of 6.5 on Richter scale. Himachal Pradesh faced a series of earthquakes in 1980, 1975 and 1905. In the year 1967 Maharashtra faced an earthquake. Assam, in the year 1950, faced an earthquake of 8.5 on Richter scale which caused the death of 1530 peoples. On 31st May 1935 Quetta (now in Pakistan) faced an earthquake. An earthquake, measured 8.4 on Richter scale, struck the Indo-Nepal border on 15th January 1934, the death toll was ten thousand. On 12th January 1885 Shillong faced an earthquake of 8.7 on Richter scale. Srinagar also faced an earthquake on 8th June 1828, measured 8.0 on Richter scale. The most devastating earthquake, by which three lakh peoples lost their lives, appeared in Calcutta on the October 11th 1787. Major earthquakes of India and world in the recent past are shown in Table 1.1 and Table 1.2 respectively.

Table 1.1 Major earthquakes of India

Date	Place	Intensity on Richter scale	Death Toll
11th Oct. 1787	Calcutta		3 Lkhs
1st Sept 1803	Mathura & Kumaon	8.5	about 300
16th June 1819	Katcha	8.0	
8th June 1828	Srinagar	8.0	1000
12th Jan. 1885	Shillong	8.7	
30th May 1885	Srinagar	7.0	
10th June 1889	Assam, Kachar	7.5	
4th April 1905	Kangra, Himachal Pradesh	8.0	20000

Date	Place	Intensity on Richter scale	Death Toll
10th Feb 1906	Kumaon and Garhwal	7.0	
8th Jul 1918	Shreemangal	7.8	
2nd Jul 1930	Dhubri, Assam	7.1	
15th Jan. 1934	Indo-Nepal Boarder	8.4	10700
31st May 1935	Quetta	7.5	60,000
14th March 1938	Madhya Pradesh	8.5	
28th June 1941	Andman	8.1	
11th Dec. 1967	Koyana	6.5	
19th Jan. 1975	Kinnaur, Lahol	7.5	
13th April 1980	Bhadranchal	8.5	
27th June 1980	Indo-Nepal Border	8.1	
10th Dec. 1987	Maharashtra	6.5	
20th Aug. 1988	Bihar	8.5	1000
20th Oct. 1991	Garhwal	6.1	1500
30th Sept. 1993	Maharashtra	6.3	35000

Table 1.2 Major earthquakes of the World

Date	Place/Nation	Intensity on Richter scale	Death toll
28th Dec. 1908	Messina, Italy	7.5	1,20,000
16th Dec. 1920	Karsu, China	8.5	1,80,000
1st Sept. 1923	Kwanto, Japan	8.3	1,43,000
26th Dec. 1932	Kansu, Japan	7.6	70,000
31st May 1935	Quetta, India	7.5	60,000
24th Jan. 1939	Chillan, Chile	7.75	30,000
31st May 1970	Peru	7.8	66,000
27th Jul 1976	Tangshan, China	7.6	7,00,000
7th Dec. 1988	Armenia, formerly USSR	7.3	45,000
30th Sept. 1993	Latur, India	6.3	35,000

FEW DEFINITIONS

Origin point of earthquake which is lying below the earth surface is known as focus of the earthquake. Point just above the focus of earthquake on the earth surface is known as epicentre (Fig. 1.2). The impact (devastation) of earthquake mainly depends upon the magnitude, distance from epicentre and the depth of focus of earthquake. Earthquakes have high depth of focus of earthquake are more dangerous. Epicentre is the centre and origin point of earthquake waves occurring on the earth surface. First of all earthquake waves struck this point, the magnitude of earthquake at this point is very high which causes major damage to the places surrounding this point. There are three types of waves originated during the appearance of any earthquake. These are termed as 'P', 'S' and 'L' waves.

P waves are longitudinal, travel with a speed between 8 and 14 km. per second, always travel below the earth surface. The speed of these waves remained unaffected when passes through solid section but slows down on passing through the liquid phase.

S waves also travel below the earth surface, these travel with a speed of 4 and 6 km. per second. These waves can travel easily through the solid sections but looses their identity when passing through liquid portions.

L waves always travel over the earth surface. These waves travel with a speed of 3 to 4 Km. per second. L waves are most destructive and devastating. These always travel with the same speed when passing through solid and liquid phases.

In the year 1935 Dr. Charles Richter of California Institute of Technology developed a scale for measuring the magnitude of earthquake. This scale is known as Richter scale. Energy released per point increase in Richter scale is thirty times greater. Intensity of earthquakes and maximum magnitude on Richter scale is shown in Table 1.3.

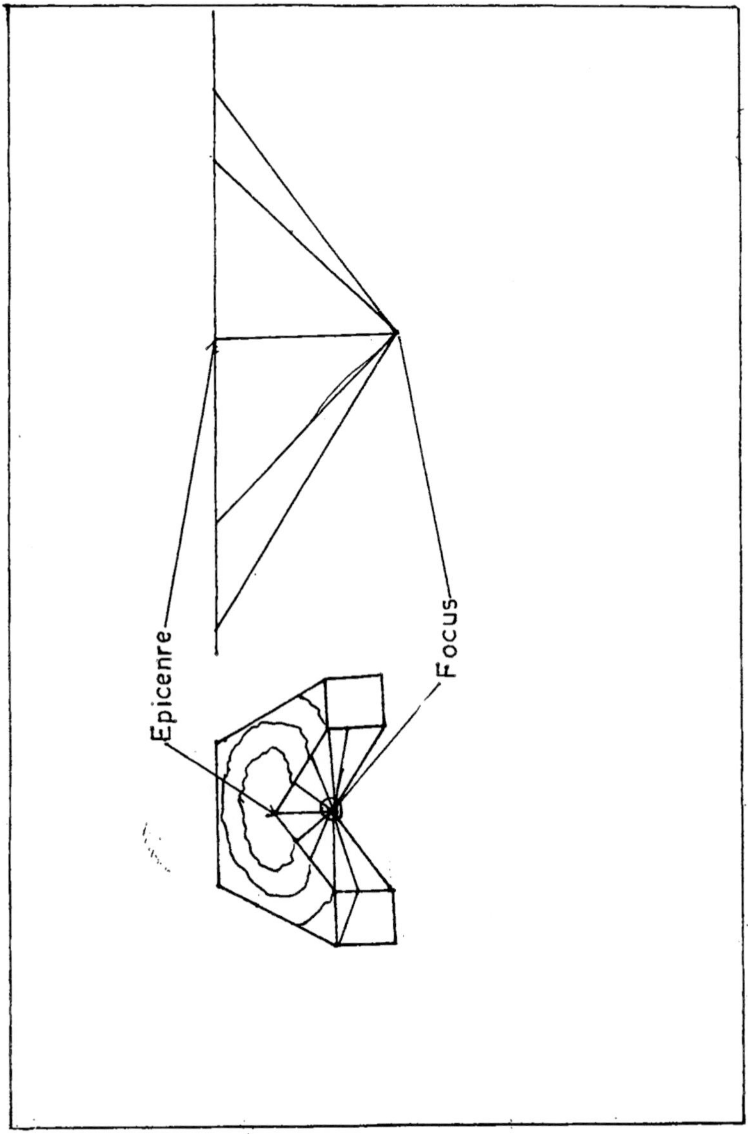

Fig. 1.2 Epicentre & Focus of Earthquakes

Table 1.3 Intensity of earthquake and maximum magnitude on Richter scale

	Intensity	Maximum magnitude on Richter scale	Impact on earth surface
I	Only measurable	below 3.5	Only measurable on seismographs
II	Very weak	3.5	Recognised by very sensitive persons or standing on the first floor.
III	Weak	4.2	Persons sleeping could feel
IV	Moderate	4.8	Most of the persons feel it
V	Intensive	4.9-5.4	Every person feel it
VI	Highly Intensive	5.5-6.1	Cracks may appear on the walls of houses
VII	Destructive	6.9	May also affect the foundation of walls
VIII	Dangerous	7.0-7.3	Cause damage to buildings, bridges, rail tracks, roads etc.
IX	Very dangerous	7.4-8.1	All construction work get dangerous collapsed
X	Most devastating	8.1-9.0	Situation of complete devastating destruction may arise

②

Understanding Earthquakes

— Y.M. Garg

'The poetry of earth is never dead....

The poetry of earth is closing never.'

— John Keats

Earthquakes were considered, in the past, an expression of anger or displeasure, wrath or curse of the Gods on humanity. Now, they are considered to be simple geological phenomena. Earthquakes are as much a part of man's environment as say, air, water, soil, or rain. Over one million earthquakes occur every year, or almost two every minutes. But most of them are too weak to be noticed. Since the beginning of the eighteenth century, earthquakes have taken a death toll of nearly three million.

Scientific study of earthquakes began around the middle of the eighteenth century, after a series of destructive earthquakes in England in 1750, and a major earthquake in Lisbon in 1755. And thus were laid the foundations of Seismology, the science of earthquakes.

Two other major earthquakes occurred in 1964, one in Niigata and the other in Alaska. Since then interest in earthquakes has increased manifold, in their causes and prediction and design of earthquake-resistant structures. Earthquake engineers now use vibration theory, material dynamics, structural engineering and geophysics in their study of earthquakes.

Stated simply, 'an earthquake is a vehement shake of the earth from natural causes'. Technically, an earthquake is a phenomenon of strong vibrations occurring on the ground, consequent to release of large amount of energy within a short period of time because of some disturbance in the earth's crust or in the upper part of the mantle.

Energy, according to the elastic rebound theory, is stored, in the first instance, because of accumulation of elastic strain on account of rupture or slippage alone a pre-existing 'fault'. In geology, a fault means the surface of a rock along which a rock has moved by a few centimeters to a few kilometres. The maximum amount of energy released in one earthquake is estimated to be 5×10^{25} ergs, about 1/1000 of the total energy released by the earth from its surface in the form of heat in a year.

Energy released is propagated by seismic waves, called 'P' waves, 'S' waves, and 'L' waves. 'P' waves are longitudinal waves, and create a 'push-pull" effect on the rock particles. 'S' waves are transverse waves and cause earth to move at right angles to the direction of the wave. 'L' the waves propagate energy along surfaces or interfaces, and set up a sort of circular vertical motion. As 'P' waves and 'S' waves do not reach the surface at the same time, an earthquake starts with faint tremors for a brief period, followed by severe vibrations for some time thereafter. 'L' waves on sea floor cause seawaves on the surface, called tsunanmis. They rise to 100 feet or more and cause damage when they break on habitated cost.

Alternatively, some scientists explain earthquakes in terms of 'Plate Tectonics' model of the earth. They believe that some 2,000 million years ago all continents were joined in one large land mass, called 'Gondwanaland', or 'Pangea' as called by Wagner in his theory of 'continental drift'. These scientists believe that the earth's surface now consists of 'blocks or plates', 100 km thick, each constituting a part or whole of individual continents and oceans. These plates are the most rigid part of the earth and are driven by convection currents by an underlying layer several hundred kilometres thick, called 'asthenosphere'. These plates jostle against one another in a complex way, getting compressed, or pulled apart, and in doing so cause earthquakes. The locations of earthquakes in the past seem to confirm the explanation of earthquakes in terms of Plate Tectonics model of the earth.

Earthquake prediction is making slow but steady progress. One approach is to predict earthquakes on the basis of changes believed or known or known to precede an earthquake. Such earthquake precursors include abnormal tilting of ground; change in strain in rock; dilatancy of rocks, which could be measured by a change in velocities, electric restivity; ground and water levels; sharp changes in pressure; unusual lights in the sky. The behaviour of some animals is also believed to undergo a distinct change prior to an earthquake. Some lower creatures are perhaps more sensitive to sound and vibrations than humans; or endowed with what one may call prescience. Another approach is to estimate the probabilistic occurrence of an earthquake statistically by relating the past occurrences to weather conditions, volcanic activity and tidal forces.

It is, however, a moot point among scientists whether a precise earthquake prediction and warning system can be developed and put to any effective use. Any meaningful prediction would need to specify the exact location and time. On the contrary, even a slight error in prediction would create more problems than it may solve. Moreover, while the knowledge may minimise casualties, this would result in social and economic confusion. Recent attempts at earthquake prediction in Japan have resulted in great tension and damage to the local economy. On the other hand, a statement like 'this bridge is likely to experience ten shocks of intensity seven or more', would serve the purpose better. The engineers can then design the bridge to withstand the likely shocks.

Seismology has found another use. Seismologists are being increasingly employed in detection of nuclear explosions and oil prospecting. Nuclear blasts release vast energy. A nuclear bomb fifty times more powerful than the Hiroshima bomb, releases energy equivalent to an earthquake of a magnitude of six on the Richter Scale. On one hand, seismology is helping in detection of clandestine underground tests, and on the other underground nuclear explosions are helping Seismological research by simulating earthquake conditions. Oil companies have used seismology for oil prospecting, by means of a study of the behaviour of seismic waves through underground strata. This eliminates the need for expensive drilling through hundreds of metres of rock.

Courtesy: The Hindustan Times

③

After Earthquakes

When an earthquake struck any region, a lot of problems like assessment of casualties and damage, problem in rescue operation, spreading of diseases, scarcity of clean water, removal of dead bodies and debris, cremation of dead bodies etc., have arrived. Few of these are being discussed hereunder.

CASUALTIES AND DAMAGE

An earthquake could cause damage to living creatures, ecology, environment and man-made structures. The effect may vary from a very little to a very large extent depending upon the magnitude of earthquake and strength of building structures. For understanding the effect of earthquake, forthcoming chapters are self explanatory.

RELIEF AND RESCUE OPERATIONS

These two tasks are of major concern after the occurrence of an earthquake in any part of the world. The intensity of an earthquake is mainly responsible for the destruction. If earthquake has high intensity magnitude, damage done is also of high order. During such operations various factors/activities affect the pace adversely.

The magnitude of damage and casualties always affect the relief and rescue operations. Bad weather also exert profound affect on the pace of such operations, it had experienced during the earthquakes of 20th October 1991 and 30th September 1993. Communication network is another factor which plays an important role during the relief and rescue operations. Damaged communication system hampers the pace of work, due to which severity of earthquake may increased manyfold, as had occurred during the 20th October 1991 in Garhwal and 30th September 1993 in Maharashtra. Bad approach routes also affect relief and rescue operations adversely, as experienced in Garhwal during October 1991. Figure 3.1 explains how various factors affect the pace of relief and rescue operations.

RECOVERY AND CREMATION OF DEAD BODIES

After any earthquake searching of dead bodies is another major task. On searching the cremation of dead bodies is also a difficult problem. Cremation of such a large number of dead bodies with rituals is very difficult, rescue workers always faced a great difficulty during this work due to the oppose of other family members if alive. During the recent earthquake of 30th Sept. 93 rescue workers faced such kind of problem. Due to the unavoidability of fuel or any other reasons if dead bodies are dumped directly into the rivers or any other water bodies cause another problem of water pollution.

DISEASES (EPIDEMICS)

If dead bodies remained unrecovered and not cremated properly, may cause various diseases which are easily communicable like cholera, dysentry, typhoid etc. on mass scale. If diseases spread and not protected can induce another series of death. Alongwith this choked drains, shortage of clean water may also aggravate the situation. Thus it is necessary to take appropriate protection measures during such calamities. Factors which are responsible for the spread of diseases are shown in Fig. 3.2.

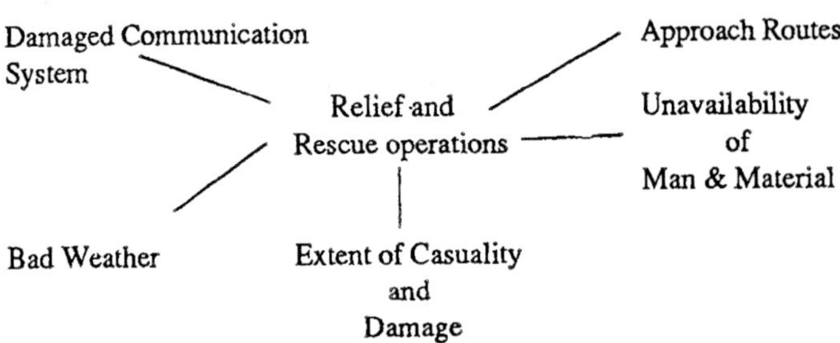

Fig. 3.1 Factors affecting relief and rescue operations

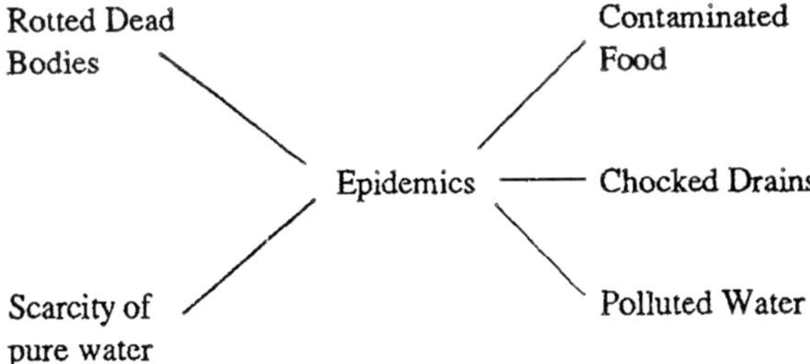

Fig. 3.2 Factors creating the problem of diseases

④
Earthquake of 20th October 1991

The Garhwal Himalaya has experienced earthquakes of varying magnitude in the past. But, after the earthquake of 1828 (which was measured 7.61 on Richter scale) Garhwal has not experienced a major earthquake before the 20th October 1991 (which was measured 6.1 on Richter scale). Seismically, the Garhwal region is a highly active zone, as two major thrusts namely Main Central Thrust (MCT) and Main Boundry Thrust (MBT) are passing across the region alongwith many other Thrusts and Tear Faults like Nayar Thrust (NT), Dunda Thrust (DT) Srinagar Thrust (ST), Gadolia Tear Fault, Tehri Tear Fault etc. These all are quite active.

That was the night of 19th October 1991 (02 53 hrs of 20th October 1991), people of Garhwal alongwith other Indians were under sweet dreams of happy tomorrow. Unfortunately, tremor struck the region and lasted about 50 seconds and washed away all the dreams. Among the five districts of Garhwal, quake affected districts are Uttarkashi, Tehri and Chamoli, the former one suffered harshly. The earthquake felt to appear along the Main Central Thrust (MCT) due to its reactivation, epicentre of which was somewhere in the vicinity of Agoda of Uttarkashi district. Shocks of earthquake were also experienced in many other parts of the country and world. The area which was affected by earthquake is shown in Fig. 4.1.

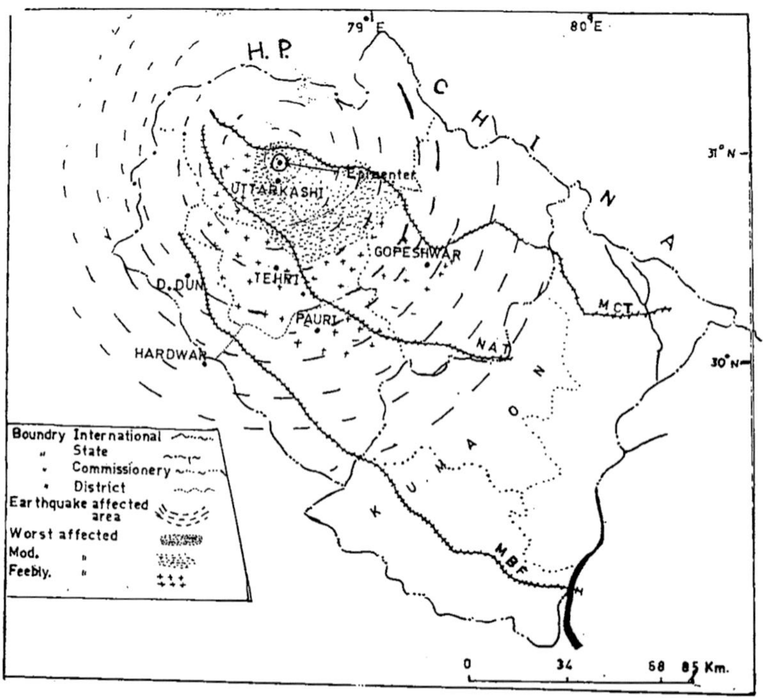

Fig. 4.1 Earthquake Affected Parts

4.1 CASUALTY AND DAMAGE

The earthquake of 20th October 1991 ravaged both life and natural resources of Garhwal Himalaya, by which near about 1,500 peoples (unofficially 3,000) lost their lives, 72 Kms of road and nearabout 42,000 houses were reported damaged. An estimation revealed that the damage caused by earthquake in Garhwal is of about 300 crores of rupees. This could be understand easily by the perusal of three news articles dated 20th October, 23rd October and 28th October 1991.

4.1.1 500 FEARED KILLED IN U.P. QUAKE

Uttarkashi hit most severely

Oct. 21.91
M.A. Hafiz
Boulders Block R. Bhagirathi

Over 200 persons are feared to have been killed in the most devastating earthquake that rocked all the light hill districts as well as the adjoining districts of Uttar Pradesh in the pre-dawn hours of today. The death toll could rise with the restoration of roads and telecommunication which have been totally disrupted following massive landslides caused by the powerful earthquake measured 6.1 on the Richter Scale.

The District Magistrate of Uttarkashi had earlier informed the State Government that he feared at least 200 casualties in his district alone where 400 villages have been flattered by the huge boulders falling from the hills.

More than 500 people were feared killed and over 3,000 injured in Uttar Pradesh, states PTI.

At least 274 died in house collapses and landslides in Uttarkashi district alone, official sources said.

Prime Minister P.V. Narasimha Rao, is likely to visit earthquake-affected areas on Tuesday. During his day-long visit, the Prime Minister will meet people hit by the devastating quake.

Uttarkashi Police Superintendent lost his 14-year-old son who is reported to have been crushed by a huge boulder that fell on the room of the bungalow where the boy was sleeping.

The earthquake totally snapped telecommunication of several parts of the Garhwal division consisting of five hill districts. There was no news of any damage from Almora and Nainital but Vice-President Shankar Dayal Sharma who had gone to the picturesque Kausani in Almora yesterday is reported to be safe.

The first tremors of the earthquake were felt at 2.53 this morning and lasted about 50 seconds. The earthquake was so powerful that the tremors were felt in Shahjahanpur, Bareilly and parts of the State capital.

According to official reports, the earthquake affected all the eight hill districts. However, Uttarkashi was the worst-hit district from where incomplete reports of casualties and damage have been received by the Government here.

In all the affected districts army has been called out to assist the civil authorities in rescue and relief operations. Besides, several companies of the Indo-Tibetan Border Police have also been deployed in rescue and relief work. The Border Roads Organisation has been asked to quickly clear the roads for faster transportation of relief materials.

Bhatwari and Gawana towns in Uttarkashi were largely damaged and it is feared that the twin towns might have lost most of its inhabitants who have been trapped inside their homes. The famous Gauri Kund in Chamoli was filled with boulders which caused death of two policemen and injuries to 15 others.

However, the Badrinath and Kedarnath temples remained unaffected. But three pilgrims were killed near Kedarnath due to landslides. According to an earlier report, 25 buildings in Uttarkashi town were devastated by the earthquake.

However, Revenue Minister Brahm Dutt Diwivedi told newsmen this evening that the death of 200 persons in Uttarkashi alone was "lacking official confirmation". He said he had been officially informed during the day that 60 persons were killed in Uttarkashi and the casualty figure was rapidly rising with more bodies being extricated from the falling debris.

A power house in Chamoli district was badly damaged following which supply of electricity to most parts of the town remained affected. The Bhagirathi river which assumes the name of the Ganga in the plains has also been blocked at one point by the heavy boulders that fell into the river due to massive landslides.

Official reports reaching here said that if the blockage in the river is not immediately cleared, there is serious apprehension of floods affecting Uttarkashi.

The official reports said that over 2500 villages in the hill districts had been affected by the earthquake and at least 90 per cent of population of these districts had been rendered homeless.

Besides three pilgrims, 17 others were also killed near the famous shrine of Kedarnath in the Himalayas. There was no report about the Tehri Dam which is under construction after environmentalists and several known seismologists had expressed their apprehensions about it being a soft target of an earthquake.

Chief Minister Kalyan Singh will be flying to Garhwal tomorrow to make an on-the-spot assessment of large scale devastation caused by the earthquake. He expressed profound grief over the loss of life and property, and conveyed his heart-felt sympathy to the families affected by the natural calamity. The Chief Minister also directed the concerned officers to launch rescue and relief operations in all the affected hill districts on a war-footing.

The State Government has arranged four army helicopters to fly foodgrains, medicines and other relief materials to earthquake-devastated Uttarkashi and Chamoli.

Besides, one State Government helicopter flew out to Dehradun today to assist the local authorities in the worst-affected areas. According to Revenue Minister Brahm Dutt Diwivedi, five hundred tonne of rice, five hundred tonne of wheat flour, ten tonne of sugar and one tanker of kerosene are being flown by the five helicopters.

The Government had also opened a relief centre at Dehradun where senior officers of the Government would be coordinating rescue and relief work. The Government has also directed a team of medical officers to reach the earthquake-affected towns and villages by tomorrow.

The Government has also formed a team of senior geologists who will be flown by a helicopter tomorrow to suggest quick clearance of a part of the Bhagirathi river filled with heavy boulders. If the natural flow of the river is not restored quickly, it may lead to floods in Chamoli.

Fig. 4.1.1 Uttarkashi: Damaged SBI Building

The quake was felt as far north as Jammu and induced cracks in some buildings in Himachal Pradesh and Delhi. In Uttarkashi the District Magistrate's house and so also that of the Superintendent of Police and the tourist home were badly damaged.

Mr. Diwivedi said that about 200 pilgrims were trapped at Gangotri in Chamoli district. As all the roads have been blocked by landslides and falling boulders the pilgrims may be airlifted later.

PTI adds from Chandigarh: Many houses were damaged and cracks developed in several buildings as an earthquake of high intensity shook Punjab, Haryana, Himachal Pradesh and Union Territory of Chandigarh early this morning.

The tremors plunged into darkness several parts of Himachal Pradesh following major power breakdown, which lasted for more than three hours, according to reports reaching here.

No loss of life was reported from any part of the region, while minor damages to buildings was reported at some places.

Doors and windows rattled with quavering sound and the shrieks of frightened people broke the mid-night silence. People ran out of their houses to safer places in the cold in many parts of Himachal Pradesh.

According to AP, in the Nepalese Capital of Kathmandu, Chief Government Seismologist M.R. Pandey said about 16 aftershocks were felt in western Nepal, but there were no immediate reports of casualties in lightly populated area. He said none of the aftershocks exceeded 3.5 on the Richter scale.

The quake was not felt in Kathmandu.

A reading of six on the Richter scale constitute a severe earthquake, and a quake of magnitude of 7 or higher is considered major and capable of wide-spread, heavy damage in populated areas.

The Richter scale is a measure of ground motion as recorded on Seismographs. Every increase of one number means a tenfold increase in magnitude. Thus a reading of 7 reflects an earthquake 10 times stronger than one of 6.

A DPA report said the US Seismology Centre in Golden, Colorado, gave the strength of the Indian quake as 7.1 points.

The official Iranian news agency IRNA said a temblor with a strength of 4.2 points was also felt in Northern Iran, with the epicentre about 300 kilometres northeast of Teheran.

Courtesy: The Hindustan Times

4.1.2 QUAKE TOLL 1,500; RS. 80 LAKH RELIEF FROM P.M. FUND

— Soumya K. Ghosh

Matali (Uttar Kashi), Oct. 23 — Prime Minister Narasimha Rao Today handed over a cheque for Rs. 80 lakh to UP Chief Minister Kalyan Singh for relief and rehabilitation work in the earthquake ravaged district.

Addressing a hastily scheduled Press conference at an Indo-Tibetan Border Police cantonment, adjacent to helipad, Mr. Rao said the money was released from the Prime Minister's relief fund. In addition, Rs. 20 lakh was also given to the Red Cross for their relief work in Uttarkashi.

(Meanwhile, the toll in Sunday's earthquake in the Garhwal hills of Uttar Pradesh has mounted to 1,500. According to reports from worst affected Uttarkashi district, says UNI.)

After an aerial survey of Bhagirathi valley, the Prime Minister said, "The massive calamity and magnitude of damages called for immediate steps. The relief and rehabilitation process has to be done with a firm footing and in a big way." Mr Rao said the State Government has tried its best to provide everything while the Central Government will assist it in every possible manner.

Fig. 4.1.2 Uttarkashi (Jamak) : Quake Disaster

"The earthquake came so suddenly and the extent of damages are so widespread that it will take some time to fully assess the situation. However, the Central Government will do its best to help the survivors and families who lost their members," he said.

Mr. Rao said the Central Government would spare no efforts to ensure that essential items like blankets, winter clothes, food items and tents be provided to affected people. Special emphasis would be made to reconstruct schools and educational institutions so that the students may attend them at the earliest, the Prime Minister added.

Stating that the grant from the PM's relief fund is "one phase of relief work", Mr. Rao said the UP Chief Minister has been asked to come to the Capital to interact with the high level Central Government committee in order to envisage a total relief package plan. He added that the Central team had already surveyed the area and some more senior officials will be here soon.

Seriously considering the fact that winter is fast approaching, the Prime Minister said, "our prime concern now is to provide shelter and proper clothes to the people of affected areas". He added that within a day or two, the Government will be in a position to ensure the extent of damage, casualty and toll.

Present at the Press conference were UP Chief Minister Kalyan Singh, former Union Minister N.D. Tiwari and H.K.L. Bhagat, AICC office-bearer Mahesh Prasad and UP Governor B. Satyanarain Reddy. Conspicuous by their absence were the area's MPs and MLAs.

Mr. Kalyan Singh said an ex-gratia payment of Rs. 20,000 will be made to families of each victim while Rs. 5,000 will be given to persons seriously injured and Rs. 2,000 to the ones with minor injury. To facilitate the rehabilitation work, the Chief Minister said Rs. 15,000 will be given to persons whose houses were totally destroyed and Rs. 5,000 to those with partially damaged houses.

Earlier, the Prime Minister had surveyed the damage in Uttarkashi town alongwith large entourage of senior State Government officials. He visited the house of the Superintendent of Police which was badly damaged (the SP lost his 15-year-old son in the house collapse), Vishwanath Mandir, Gopal Mandir, the ladies of Government Inter-college and the district hospital.

At the hospital, Mr. Rao met the injured persons and later asked the Chief Medical Officer and a team of doctors about the condition of patients. Several injured narrated their harrowing experiences and even complained of lack of amenities. The Prime Minister openly rebuked

Commissioner Nagendra Singh after several injured and their family members complained of inadequate relief work. Mr. Rao also held a discussion with relief workers including RSS-VHP workers.

The bulk of relief work is being done by several non-governmental organisations like Doon School, Alumni, Bharat Electronics Limited, Kotdwar, Divine Life Society, Degree college students, several ashram members of Rishikesh, IAS probationers and PWD Mahila Samiti.

Courtesy : The Hindustan Times

4.1.3 CRACKS IN UTTARKASHI HILLS ALARMING

— Soumya K. Ghosh

Sangrali (Uttarkashi), Oct. 28 — Environment scientists and rock specialists are seriously concerned about the wide cracks which had developed on the mountains, perched right above the Uttarkashi township. Though opinions differ regarding the reason of the crack, the situation here is alarming.

Situated 7 km above Uttarkashi, Sangrali is witnessing an exodus everyday as hamlets are being deserted. People are trekking down hill to safer grounds, despite repeated assurances by the district administration. Down in Uttarkashi, these panic-stricken people are settling wherever they are finding open spaces.

However, as more reports poured in from interior areas, a similar situation had arisen in Agora, which was the epicentre of Oct. 20 earthquake, and Bhankoli. People of Kelsu, Laksheshwar, Tekhla, Gangori, Rawara, Kalyani, Sangam Chatti, Seku, Gajouli, Naongaon, Bhankoli and Agora are walking downhill for settling down.

The cliff overlooking Agora, it is reported, is just dangling, with loose earth constantly falling down on the village. Pyar Singh from Agora said: "After the Sunday earthquake, the ground beneath the village had caved in. Moreover tremors are taking place everyday. We in the mountains believe that once a rock is loose, it is bound to fall. We are appealing to the district magistrate to remove us from here".

Fig. 4.1.3 Damaged House

Already, houses in Agora and Bhankoli have developed wide cracks and would fall like a pack of cards if the tremors continued. The locals are complaining that they would get a meagre Rs. 5,000 for reconstruction of their houses, as against Rs. 20,000 for completely destroyed ones, from the State Government. Malti Devi, a Bhankoli housewife, said: "In another few months, our homes would crumble and fall because the cracks are so wide. In fact, we are sleeping in the open lest the house should fall down. Just because our houses have been branded partially damaged, we will get less money".

In areas of Kelsu, Gajouli, Bhakoli, Dansa, Dhandalka, Seku, Salu, Saura and Sari, residents pleaded the Army and ITBP rescue and relief team, to break down the houses, after bodies were extricated, in order to attract the Rs. 20,000 aid. A senior ITBP officer, commanding a relief team, said in Seku that some time we did that so that these "poor people could get some benefit".

Meanwhile, the district administration at the Uttarkashi headquarters is routing all relief materials and supply of essential commodities to remote and high altitude areas, through the ITBP and Army. Every morning, fresh contingents of jawans are seen walking towards the mountains from the Gangori, Gawana, Netala, Hina, Mahidanda and Rawara sides. They are assisted by medical and geological survey teams.

A controversy has arisen regarding the death toll, when village-wise headcount was taken up. While figures are pouring in from various corners of the mountain district, there is wide difference with the tally of the officially declared death toll, which is 680. It is estimated that the number of deaths due to the Oct. 20 quake was well past 1,600.

Till Oct. 22, rescue teams and Army helicopters were bringing bodies from different villages to Uttarkashi district hospital. Sensing trouble, district magistrate Sunil Agarwal issued orders that the bodies should not be brought to Uttarkashi and instead be cremated locally. This is surmised to be the reason why the death toll was "contained".

Following the Oct. 20 quake, there had been a series of landslides in which a large number of people had died in the last four days. Though an official count could not be furnished, it is estimated that over 150 people, including women and children had died.

Meanwhile, in a cleaning-up operation, 45 bodies which were Floating at the sluice gates of the Maneri dam, were removed by the district administration. It is reported that several bodies floated downstream while Jamak sources said due to paucity of firewood for cremation, many bodies were thrown on the dam waters.

According to an ITBP official posted at Matli, there had been extensive damage in Pilang (7,000 ft.), Bung, Alet, Pata, Manpur, Pow and Kishanpur areas, where despite several helicopter sorties, relief material could not reach. Several teams were sent to these far-flung, high altitude places with supplies and relief measures. A total of 500 jawans, headed by a dozen commanding officers, drawn from Matli (18 Battallion) and Mahidanda (4 Battallion) have been already sent to these places.

Aiding the relief operations apart from the ITBP and Indian Army are the Border Security Force (BSF), Provincial Armed Constabulary (PAC), Central Reserve Police Force (CRPF) and Special Service Bureau (SSB).

Fig. 4.1.4 Maneri : Dam Site

The helicopter services, provided by the Sarsawa and Dehradun airbases had stopped from Saturday as district administration officials said that the "situation is under control". Only a few sorties are being made everyday from Matli helipad for VVIP aerial surveys. Heading the chopper support were Wing Commander Kapoor of Sarsawa base for Chetak and Wing Commander Dahiya from Dehradun base for MI 8.

Meanwhile, ITBP Director General D.V.L.N. Ramakrishna Rao, while inspecting the quake-ravaged valley, said two affected villages have been "adopted" by the ITBP for reconstruction as well as employment opportunities for eligible youth.

Courtesy : The Hindustan Times

4.2 RELIEF AND RESCUE OPERATION

Relief and rescue operation to the quake victims started just after the first news of havoc. It was slow in the beginning due to non availability of material and unapproachability of the area but later on picked up speed when things were made available. Indian Army and Paramilitary Forces played a significant role in this operation. Prime Minister (PM) of India during his visit to the quake affected area donate Rupees 80 lakhs for relief operations. The Chief Minister (CM) of Uttar Pradesh (UP) announced that Rs. 20,000 will be made available to the families of each victim and Rs. 5,000 will be given to the persons seriously injured, while Rs. 2,000 to those with minor injuries. He also announced Rs. 20,000 for fully damaged house and Rs. 5,000 for partly damaged house. Besides this the victims were also given Medicines, Food and Temporary shelter to fight against cold and hunger. The Garhwal had been bestowed sympathy from all over the world in the form of money and relief material.

People of interior region were reported to have less amount of relief material, which had happened due to the non approachability of that section caused by the destruction of roads and foot paths. Unplanned distribution is also a causative factor for such misshapening in the region. Concerning the relief and rescue operation few forthcoming articles are self explanatory.

4.2.1 MEDICARE PLAN FOR EARTHQUAKE VICTIMS

New Delhi, Oct. 29 — The Loknayak Jayaprakash Narain hospital authorities have drawn an elaborate plan to provide medical care to the victims of recent earthquake in Uttarkashi. The plan taken particular care of the survivors who will be faced with the onsetting winter without a roof overhead.

According to hospital sources, a three-member team of hospital doctors is leaving here for Uttarkashi on November 6 to make an on-the-spot assessment of the situation in terms of the medical requirements. The team will operate with the co-operation of the local political leaders and medical experts.

As per the plan, the disaster ward of the hospital has been thrown open for the earthquake victims who need multi-disciplinary care of specialised nature like for crush injury and multiple trauma. A disaster relief cell has also been set up in the casualty section of the hospital which will remain open for 24 hours and monitor the condition of the quake victims admitted there.

Fig 4.2.1 Damaged House and Relief

The hospital authorities have also launched a campaign to collect relief materials and supply them to the victims on the spot. These materials include medicines, dressing material, blankets and woollens.

The team of doctors, which will visit the quake-hit areas, include Dr Anil Gurtoo, Dr. Shinoy Robinson and Dr. I.P.S. Chabra. They will work in the area for three to four days.

Courtesy : The Hindustan Times

4.2.2 UP DENIES BUNGLING IN RELIEF

Lucknow, Oct. 30 — The State Government today denied any bungling in the distribution of relief materials to the earthquake-affected people in the three hill districts of Uttar Pradesh and asserted that distribution of relief had been stepped up this week.

An official spokesman told newsmen here this evening that the State Government had sanctioned a sum of Rs. 62 crore for relief under various heads which include Rs. 35 crore as subsidy for building houses. He said a total of 28,220 houses had been fully damaged while the number of the partially damaged houses is 20,658. A sum of Rs. 20,000 is being given for fully damaged houses, he said.

For partially damaged houses, a sum of Rs. 9.16 crore had been placed at the disposal of the district collectors of Uttarkashi, Tehri, Garhwal and Chamoli. He also denied allegations that one blanket per family was being given by the State Government and said that five blankets per family were being given.

Fig 4.2.2 Helicopters for Relief and Rescue operation

The spokesman denied that 3,000 persons were killed by the devastating quake that rocked Uttarkashi, Tehri and Chamoli on Oct. 20 and asserted that only 786 bodies had been found so far. He, however, admitted that the official death figures related to body counts only. "The figure may swell when more bodies are extricated under the debris of the collapsed houses", he said.

On the other hand, leader of the Opposition in the State Assembly and UP Janata Dal president (Ajit faction) Kailash Nath Singh Yadav complained that the relief committees in the three hill districts had been packed with the members of the BJP even though the earthquake was a national calamity.

On his return from the quake-devastated districts where he inspected the relief camps and met a large number of victims, Mr. Yadav complained that the official relief was too meagre and not reaching the far-flung villages of Uttarkashi.

Talking to newsmen, Mr. Yadav claimed the devastating earthquake had killed at least 3,000 persons while the local officials believed in "suppressing" the casualty figures to minimise the gravity of the calamity. He said most of those whose houses had been destroyed were still living under the open sky due to acute shortage of tents.

Mr. Yadav urged the Chief Minister to form an all-party relief committee for speeding up relief work in the three districts and suggested that the compensation of Rs. 20,000 announced by the Government to the next of kin of those killed be raised to Rs. 1 lakh. He also suggested that tents and kerosene should be rushed to the three districts on a war-footing.

Rs. 62 CR FOR VICTIMS: The Uttar Pradesh Government has decided to provide Rs. 62 crore for the earthquake victims in Garhwal region adds PTI.

Chief Minister Kalyan Singh told r porters that the amount of ex-gratia to the victims has been increased.

He said Rs. 20,000 has been sanctioned for each collapsed house and Rs. 5,000 for damaged houses.

Ex-gratia amount between Rs. 20,000 and Rs. 60,000 has been fixed for family members of those killed in the earthquake.

Mr. Singh said, according to his estimate, the State had received an assistance of Rs. 80 lakh only out of Rs. 70 crore sanctioned by the Centre for quake relief as against a loss of about Rs. 300 crore incurred.

The Chief Minister urged the Centre to provide the maximum assistance to meet the casualty.

The Union Government proposes to spend Rs. 40,000 crore for expansion and modernisation of the telecommunications network during the Eighth Plan period, Union Deputy Communications Minister Rangaiah Naidu said in Bangalore on Monday.

Courtesy : The Hindustan Times

4.2.3 DIRECTIVE TO UP ON QUAKE REPORT

New Delhi, Nov. 3 — Union Agriculture Minister Balram Jakhar today directed the Uttar Pradesh Government to submit a project report immediately for improvement and development of infrastructure in the five earthquake affected district of Garhwal division with the World Bank assistance.

An official release said Mr. Jakhar visited the villages between Uttarkashi and Harsil today which were badly hit by the recent earthquake and oversaw the relief operations.

He also visited the injured in the hospitals and had discussions with zila pramukhs, sarpanches and officials involved in the relief work.

The Agriculture Minister also gave a cheque of Rs. 4 lakh to the secretary rural development of the Uttar Pradesh Government for seeds and assured a similar grant soon. Mr. Jakhar reviewed the progress of restoration work of irrigation channels, agricultural fields and inquired about the position of supply of seeds and planting materials for horticultural programme.

Meanwhile, the National Committee of the Congress for Relief to the Earthquake Sufferers today sent out fresh appeals to Chief Ministers, party MPs and other office bearers for collecting funds to help the quake victims in relief and rehabilitation, reports HTC.

The committee, which met today under the chairmanship of Mr. N.D. Tiwari, felt that steps for permanent settlement of those living in the hills in earthquake-prone areas should be speedily considered.

It decided to request the Urban Development Ministry to ask the Institute for Housing, Roorkee, to prepare suitable models. The committee on its part would strive to make available such models made on the experience of Japan and Mexico. The HUDCO should also be involved in a big way.

BADRINATH: The recent earthquake in Uttarkashi district has not damaged the famous Badrinath and Kedarnath temples but had caused cracks in some other temples and houses, UNI reports.

Mr. K.N. Pencholi, executive officer of the Shri Badrinath-Kedarnath Temple Committee, said ancient temples that had developed cracks included the Saraswati temple in Kalimah and Burha in the Mandakini Valley.

Mr. Pencholi who is supervising the relief operations in the affected areas, said the Kedarnath shrine will be closed after Deepawali while the Badrinath temple will remain open till Nov. 17.

4.2.4 QUAKE VICTIMS FEEL CHEATED

—Soumya K. Ghosh

District administration will be submitting the final assessment report to the State Government, regarding relief operations in the quake-torn region by tomorrow amidst widespread resentment among the local people who feel that the distribution system had been discriminatory and haphazard.

The State Government had deployed 25 Block Development Officers, in addition to personnel of Revenue and Supply Departments, to aid the relief distribution system and to compile a detailed assessment of destruction and relief measures. These officials who were drawn from various districts of the State were aided by the district administration teams.

However, official sources said here today that remote areas, which had been severed due to heavy landslides following the Oct. 20 earthquake, are still taking a beating, in terms of relief operations.

Areas of Netala, Harsil, Pilang, Salang, Agora, Bhankoli, Janak, Dharali and Mukhwa have been "partially" provided with relief operations. Officials said that these areas are so much in the interior and the roads leading to them are difficult to negotiate that distribution of essential commodities are next to impossible.

It is reported that in certain areas caste factors had dominated the distribution system. Locals alleged that members of scheduled caste and tribe, who had been worst affected in the quake, were denied supplies. Upper caste patwaris and tehsildars ensured that SC/ST members be the "beneficiaries of the second list". In terms of distribution, second list means those people, who were not adversely affected by the disaster.

In Dunda, where the local administration has opened a relief office, several villagers complained that they were refused to be enlisted in the list for the housing subsidy. The Government has fixed Rs. 20,000 for totally damaged houses and Rs. 5,000 for partially damaged ones.

Meanwhile, officials involved in relief operations today said that villages which are near the roads have received the maximum relief aid, while far-flung areas on higher altitude were provided less.

Mohan Tripathi and Prem Parkash, two Block Development Officers operating in Bhatwari Tehsil, said villagers of roadside hamlets despite being given the maximum aid, complaint of irregular distribution. "They had been given the most of the relief and in time. When it comes to complaining, they are the first".

Courtesy : The Hindustan Times

4.2.5 PLAN TO FIGHT COLD ON CARDS IN UTTAR-KASHI

—Soumya K. Ghosh

Uttarkashi, Dec. 5 — The Uttarkashi district administration has proposed an ambitious project called the Operation Warmth in order to combat the crippling cold of the quake-ravaged Bhagirathi valley. Official sources said today that the project will be operative within two weeks.

What is considered as the "mopping-up programme" after the relief distribution in the region had miserably failed despite abundance of relief aid and resources. The Operation Warmth will provide woollen garments, essential edibles, winter housing aids and life saving drugs.

Aiding this operation will be a few Government of India enterprises like IFFCO and Indian Oil apart from national and international voluntary agencies.

The operation will function in a similar pattern of the relief distribution which is through four tehsil headquarters of Dunda, Bhatwari, Purola and Bharkot. The Dunda and Bhatwari sub-division will have 21 and 11 sectors, each sector will have 8 to 10 villages under it, mainly those which are near the roadsides. For hamlets which are in high altitudes, district administration officials along with local support will operate.

Deputy Collector R.P. Singh said that in Bhatwari tehsil, a total of 98 villages have been earmarked for the operation. He said a large quantity of tin sheds will be also provided for high altitude villagers, while 50,000 monkey caps, mufflers and woollen socks would be also distributed.

It is learnt that after public pressure on the district administration, the Indian Army, Indo-Tibetan Border Police and the Seva Suraksha Bandhutya will be shouldering a large chunk of the operation's distribution.

In addition to this, the administration is also contemplating to provide low interest bank loans against agricultural and animal husbandary inputs for the quake-affected region. For this, quite a number of nationalised and scheduled banks have been approached. It is learnt that these rural bank branches are also in a process of arranging a list of this effect.

Meanwhile, a World Bank team made a helicopter and land survey of the devastated areas, along with senior Central and State Government officials. It is reported that the World Bank team also had soil and engineering experts. Administration officers said the WB team had come to assess the damage of roads, bridges, Government offices, hospitals, private buildings and educational institutions. A survey report and negotiations regarding the modalities of WB relief measures with the UP Government is to follow soon, though administration officers felt that such infrastructural preparations takes a little time.

There was seething resentment among the relief seeking villagers, who had thronged the Bhatwari tehsil office, from high altitude areas. Jyot Singh of Pala (where the snowfall has started) showing this correspondent peeled-off strips from the tin sheets said: "What kind of mockery is this in the name of relief. They are forcing us to buy such substandard tin which will not last the first rainfall, what to say about the snow. The Government is deducting Rs. 375 per tin sheet."

Hukum Singh, who had walked down 17 kms from Barsu said "A week ago our village was supplied with four tarpaulin sheets, though there are about 60 families in the village. These tarpaulins were heavily starched. When dew fell on it, the tarpaulin was finished. Now you can see the sky through it. They deducted Rs. 750 for each tarpaulin sheet".

Anger war writ on the faces of the villagers, when they were given relief blankets with a deduction of Rs. 155 for each, from the housing subsidy fund. Prem Singh Rawat from Sailang said even in Uttarkashi open markets these quality of blankets cost around Rs. 80. "We are poor and we need these blankets desperately now. So we have to accept it. But this is injustice", he added.

Reports are filtering in that in Bhatwari tehsil, which the administration proudly claimed to be "totally covered with relief supplies", a large number of villages have not even received the compensation for the injured, dead cattle and even ex-gratia for the next of kin of victims.

Surprisingly, Christian voluntary organisations had fanned out in almost all remote areas. Notable among them is the Discipleship Centre of Delhi, which had covered about 1,300 families of Bhatwari and Dunda block. In an organised manner, this organisation had distributed a two blanket roll along with five sweaters to each family apart from 4 tin sheets and 4 wooden sleepers to each household. In Bhatwari, they had covered Sari, Sura, Lata, Salu, Kumalty, while in Dunda, they fanned out to Digtol, Baman Gaon, Panchan Goan, Udalka, Maja Sari, Jari Dumka, Kulath, Hitanu, Danda, Najaf and Mira Gaon.

Fighting in vain with authorities at Uttarkashi Collectorate is the pradhan of Bhankoli (near the epicentre of the earthquake Agora) Tribhuvan Singh. He said "In 1956 earthquake, nearly 200 acres were badly damaged. The Government promised us to shift the village down to Rewari. A move was also taken in 1978. After the Oct. 20 quake, the villages has 30 mts deep cracks. Moreover, landslides are rampant. At this rate, the entire village will perish."

Courtesy : The Hindustan Times

4.2.6 VICTIMS SHIVER AS OFFICIALS POCKET RELIEF

—Soumya K. Ghosh

Jama (Uttarkashi), Dec. 1 — Bhardi, tears staining her wrinkled Mongoloid face sat glued in front of a heap of ruins, which once housed a happy family. After losing four minor grandchildren and two daughters-in-law, she spends most of the time near the ruins, silently weeping.

But the same woman stared back in anger when asked about Government relief: "What relief do you call this torn silk saree, tattered blue sweater and a handful of grains? They (Government) are a bunch of butchers. They have no mercy for us. At this rate, no child will survive in this village this winter".

Jamak, which suffered the most in the earthquake with a toll of 79, speaks of the quality of Government aid. A village elder, Kamal Singh, said were it not for Geeta Press and Adara (a Seventh Day Adventist Church unit) volunteers, the hamlet would have been left to its fate.

Incidentally, after the quake the village had been visited by four State Ministers, former Prime Minister Chandra Shekhar and former U.P Chief Minister N.D. Tiwari, apart from almost all senior Government officials posted in the region for relief and rehabilitation operations.

Perched atop an adjacent flat land, the Jamak survivors huddle in their makeshift tents, tarpaulin-covered sheds and haytop huts along with their women, children and cattle while a chilly wind blows over it. Drinking water has to be fetched from a spring, four kms below the village as water from the Government pipelines was refused to the survivors. A resident, Pyar Singh, complained: "The Government has not even helped me to extricate the bodies of animals from the debris. We were given tin sheets, woollen clothes, tarpaulin sheds and grains, but it is also inadequate and of such poor quality."

It is a sad sight to see hundreds of hapless survivors, waiting down at the roadside for relief blankets or tin sheets. They had to trek down 45-50 km after waiting for weeks for relief supplies.

Villages like Pati, Dwari, Baithal, Kairal, Notin, Salang, Salu, Tilang, Chaba, Duggadal, Belog, Sisba, Kujjan, Tharang and Sanglai are relatively, untouched by any major relief measures.

Waiting for the trucks carrying tin sheets, Priti Singh, a village elder from Bandrani hamlet, said there were gross anomalies in the distribution of housing subsidy. "The Government has announced payment of Rs. 20,000 for totally demolished houses, but we are getting only Rs. 8,000 in hand. They are deducting money for the tin sheets, tarpaulins, blankets and woollen garments. Is this fair? What can you do with such a paltry sum? Not even a cattle shed can be constructed with it. Moreover, the market prices of these items are far lower."

Other villagers also seconded by the pradhans also claimed that tin sheets, tarpaulin and blankets cost far less in the market. Parduman Singh an arts graduate angrily said: "The money is going into the pockets of contractors and senior Government officials. Why don't they give us the housing money to let us buy our own requirements? This is trechery".

Courtesy : The Hindustan Times

4.2.7 QUAKE VICTIMS RALLY AGAINST POOR RELIEF

New Delhi, Dec. 5 — A large number of earthquake victims from Uttarkashi participated in a rally at Boat Club here today to protest against inadequate relief and rehabilitation work in the devastated areas and to highlight the miserable plight of thousands of sufferers like them in the hills.

Shouting slogans against the Uttar Pradesh Government and the Centre for their alleged apathy and neglect, they marched in procession to Akbar Road and Tughlaq Road. They were led by Swami Rama, founder of the Earthquake Rehabilitation Committee at Uttarkashi which is carrying on relief work on a large scale.

Frustration and anger writ large upon their faces, the marching men and women rallyists waved placards condemning the neglect of the hill region of Garwhal in its hour of crisis.

Later, addressing a large gathering at the Boat Club, Swami Rama deplored that due to a political tussle between the Centre and the State Government, even a month after the tragic earthquake the shelterless survivors were suffering in the cold and due to deprivation.

The Governments must realise that it was an emergency which had to be tackled on a war footing because in a few weeks it would begin snowing in the mountains and the unfortunate survivors would freeze to death, he pointed out.

He stated that he had met Prime Minister P.V. Narasimha Rao and apprised him of the facts. He had demanded that all relief and rehabilitation work be immediately handed over to the Army and Paramilitary Forces as the government machinery had proved incapable of handling the task.

He also demanded that the dependents of those killed in the earthquake should be given at least Rs. 1 lakh and those injured be given Rs. 20,000.

There were reports of misappropriation of relief funds, he stated, adding that an all-parliamentary committee should be set up to investigate the gross misuse of funds.

There was also an urgent need to rebuild the affected houses in the hills immediately at Government expense and restore the traditional economic structure in a systematic and scientific manner, Swami Rama stressed.

The Prime Minister had assured him that steps would be taken to speed up the relief and rehabilitation measures soon. Similar assurance was also given to former Chief Minister N.D. Tiwari, who led a delegation of Uttarkashi people to meet the Prime Minister today. Mr Tiwari also addressed the gathering of the quake-affected people at the Boat Club today.

Courtesy : The Hindustan Times

4.2.8 PLEA TO INDUSTRY ON QUAKE RELIEF

New Delhi, Nov. 4 — Uttar Pradesh Tourism Minister Kedar Singh Phonia, who is one of the two State Ministers supervising relief work in areas worst-affected by the recent earthquake in the Garhwal region, today appealed to industrial houses and voluntary organisations to adopt villages for rehabilitation and reconstruction work.

Talking to newsmen here, he said the Vishwa Hindu Parishad had adopted Didsari, the Bharatiya Janata Party had adopted Netala and Rontharu village had been adopted by the Rashtriya Swayamsevak Sangh. These villages are located in Uttarkashi district.

In the three worst-affected districts, the quake played havoc in 175 villages of Uttarkashi, 250 in Chamoli and 473 of Tehri Garhwal.

The Minister denied the allegation that the State Government wanted only the ruling party BJP and its allied organisations to take part in the relief work. "I regret the politicalisation of the great human tragedy", he said.

The Minister agreed that till four to five days ago, the administration could not claim to have reached relief work in all the affected areas. Today, he said, he could say that all areas had been covered.

Mr. Phonia said while the loss of property was estimated to be Rs. 300 crore, the loss of life, and the impact the quake left on the life of the people was immense. The biggest challenge before the State Government was to save the lives of the injured and other normal people in the coming months of winter.

The State had sought Rs. 200 crore — two-third of the physical losses — from the Central Government but so far only Rs. 80 lakh had been released by the Prime Minister. Besides, the State had been provided a loan of Rs. 30 crore from the National Housing Bank, Rs. 16 crore had been released from the natural calamity fund and Rs. 25 crore were allowed in overdraft.

Irrespective of the help from the Centre, the State Government had decided to extend Rs. 20,000 help to owner of each fully damaged house. As most of the deaths took place due to heavy stones forming the ceiling, half of this help would be in the form of light building material—Rs. 7000 for wooden planks and Rs. 3000 for tin sheets. Eighteen thousand tarpaulins had been sent to the three worst-affected districts. Blankets and food supplies were among the relief material.

The main control room for the relief had been set up in Dehradun under the charge of the Meerut Commissioner.

The Minister said he did appeal to VIPs that they should not visit the worst-affected areas as this could hamper relief work. Still many leaders like former Prime Minister Chandra Shekhar, Mr. Rajesh Pilot and Mr. Farooq Abdullah visited the area. Prime Minister P.V. Narasimha Rao's visit was welcome as the State wanted Central assistance from him.

Meanwhile, the Chinese Government has donated US $ 50,000 worth of goods, including blankets, clothes, and foodstuff to India's earthquake-stricken areas.

This was conveyed to Indian Ambassador in Beijing Salman Haidar by Mr. Wang Yingfan, Director General of the Asian Department, Chinese Foreign Ministry, a Chinese Embassay news release issued today said.

He also expressed the Chinese Government's and peoples' deep sympathy and sincere solicitude to the Indian Government and the people in the affected areas as well as the bereaved families. The relief material will be sent to India immediately.

Courtesy : The Hindustan Times

4.2.9 CM REVIEWS RELIEF OPERATIONS

Kucknow, Dec. 8 — Chief Minister Kalyan Singh has directed that all the decisions taken in respect of relief measures in earthquake affected districts of the Garhwal division should be executed in a time-bound manner by December 15 next and for achieving this the administrative machinery should be geared up on war-footing.

Fig. 4.2.3 A long wait for the relief material at a Uttarkashi bus stand
Courtesy : The Hindustan Times

The Chief Minister gave these instructions while reviewing relief operations in earthquake hit areas at a high-level meeting of officers late night yesterday. The meeting was attended among others by Irrigation Minister Om Prakash Singh, chief secretary and the principal secretaries and secretaries of concerned departments besides the relief commissioner.

The Chief Minister directed that the relief operations in earthquake hit areas should be carried out with accelerated speed by making the administrative machinery more affected. Additional personnel should be deployed.

The Chief Minister has ordered that for saving the earthquake victims from cold, shelter should be provided as far as possible under a time-bound programme. Besides the tin sheets being given to the earthquake victims for this purpose, he said. Construction of community shelter camps should also be immediately started and preference be given to higher altitudes falling above 5000 feet.

It was decided to form 100 teams of engineers drawn from various construction departments to speed up the construction works. A sum of rupees four crore has been sanctioned for this purpose. In case the earthquake victims wished to come to lower heights, adequate arrangements for their temporary shelter be made accordingly.

The Chief Minister also directed that besides the quake affected district headquarters progress or distribution of relief should be regularly reviewed daily at the State headquarters. He said that he would himself review the progress or relief distribution once a week. It has also been directed that every quake-affected family should be provided at least five blankets to protect the inmates from cold. As many as 32,500 additional blankets have been provided for the purpose. A total of about 80,000 blankets have so far been distributed.

The team of senior officers sent to the quake affected areas for evaluating the relief operations from the State headquarters also presented their reports to the Chief Minister at the meeting.

A sum of Rs. 95 crore has so far been made available for relief operations in the quake-affected areas.

Courtesy : The Hindustan Times

4.2.10 QUAKE VICTIMS PRONE TO KIDNEY FAILURE

Chandigarh, Oct. 23 — Earthquake-affected residents of Uttar Pradesh have a high possibility of developing a sudden onset of kidney failure, according to the head of the Nephrology Department in PGI, Dr K.S. Chugh.

Dr. Chugh said here today that past experience showed that individuals trapped under falling masonry and collapsed multi-storeyed buildings were prone to kidney failures.

He said the international society commission on acute renal failure, based in the Mackenzee Health Science Centre, Edmonton (Canada), had conducted a study on problems of "crush injury induced acute renal failure".

The commission had, he said, offered machines, including dialysis, doctors and other medical personnel for the earthquake victims in Uttar Pradesh.

Dr. Chugh said the PGI Nephrology Department was also prepared to send a medical team of doctors, nurses and other personnel to Uttarkashi.

PUNE: Information about sudden rise or fall in the water level in wells, change in well water taste, noticeable tilt of ground, rock-falls and abnormal animal behaviour in earthquake-prone areas can help seismologists in predicting earthquakes, says a city-based senior seismologist.

Prof. B.G. Deshpande said seismologists the world over have succeeded in long range prediction of earthquakes but short range or "imminent" predictions of a few hours or days, which alone could help save human life, was not as yet possible by seismological methods only.

Prof. Deshpande, who wrote a book "Earthquakes, animals and man", said animals, birds, fish and insects respond to sudden changes in environment caused before an earthquake.

Courtesy : The Hindustan Times

4.2.11 EPIDEMIC THREAT IN UTTARKASHI

—Anil Anand

New Delhi, Dec. 6 — Having survived the tremor that shook the Uttarkashi region, the fear-stricken inhabitants of over 50 far-flung villages are facing the danger of epidemics owing to inordinate delay in relief measures.

A survey conducted by three different teams of the doctors of the Loknayak Jaiprakash Narayan Hospital has revealed that the people spending sleepless nights in the open have already been afflicted with a number of diseases. These ailments have been caused by the deadly combination of malnutrition, cold weather and the psychological trauma.

According to a report prepared by these doctors, the worst-affected are the children below the age of 15. Over 60 per cent children in this age group were reportedly suffering from respiratory tract infection. The doctors found that though this disease was common in these areas, the number of cases recorded after the earthquake was nearly four times more. The infection which was fast spreading could prove much more devastating in the months to come if immediate medical help was not rushed to these villages.

The doctors further found that the pockets around Agoda epicentre, where the tremor caused maximum damage, have an increased prevalence of gastroentritis. At least one person has fallen victim to this disease and more casualties were likely on this account as scores of others were having acute bouts. This was being caused by the intake of contaminated drinking water and absence of proper facilities for defecating.

The villages are located in the higher reaches in Dunda and Bhatwon blocks of Uttarkashi. Assisted by the Institute of Mountaineering, the doctors chose to reach these villages as the concentration of relief activities was mostly in the accessible areas around the district headquarter of Uttarkashi. Setting up two different base camps in Dhontri and Kamad, they toured Bhatiyara, Chaundiyat, Saundi, Dhikoli, Saur Laudara, Dung, Siri, Kongarh, Khamkhal, Ghorpur and Mukhmal in the first phase. In the other trek, the second team operating from Kamad, went to Thandi, Bhardhar, Sain, Branpuri and went to Medh via Ayar pass. Later they also went to Bharkot and Satyanyali.

The doctors during the course of their interaction with the survivors felt that there motivation to work for survival was completely benumbed. They were utterly fear-stricken due to loss of family members and the repeated jolts which the areas have been experiencing eversince the October earthquake, said Dr. Anil Gurtoo, pool officer and a member of the team. He said these problems on the health front were compounding due to lack of shelter, cold exposure, overcrowding under the makeshift tarpaulin dwellings and malnutrition.

He further said the productive capacity of the entire population are benumbed into a motivational paralyses. This was being further aggravated by the feelings of despair, helplessness and dependency. In reply to a question, he said such emotional tendencies if not checked through proper health care and other relief measures, would crystallise into chronic dependency which would effect their efficiency and productivity in the future.

Dr. D.D. Kulpati, the Medical Superintendent of LNJP on whose initiative these relief measures were started, said decreased in caloric intake of the villagers was another cause of worry which would strengthen the roots of malnutrition and weaken their internal immunity systems. Due to the non-availability of foodstuffs, they have considerably lowered their diet. As against normal intake of 500 grams of rice and 100 grams of pulses per head, they were now taking 200 grams of rice and 50 grams of pulses. This phenomenon was more marked in villages which were located at a height of more than 5000 feet.

He was of the opinion that these areas have been ignored due to inaccessibility and lack of media attention. He felt that organisations like the Nehru Institute of Mountaineering should be actively involved to reach these areas. The institute principal, Group Captain, A.K. Bhatacharya, not only keenly followed the efforts of the doctors but provided them all possible help in reaching these areas and shifting some of the serious patients to the hospitals and dispensaries.

Dr. Kulpati revealed that at present three earthquake survivors were under treatment in the trauma ward of the hospital which has been kept open for the tremor victims. He said a detailed report was being prepared to be submitted to the Government so that a comprehensive plan could be formulated not only to save the survivors but ensure their healthy growth.

Courtesy : The Hindustan Times

4.3 IN SEARCH OF A FUTURE

—Chand Joshi

Jamak village: The men gather on the banks of the Bhagirathi in a gesture of cathartic repentance. Every one of them was getting his head shaved as a ritual for the 73 villagers who had died in the earthquake a month ago. There had been no cremations, no last rites—merely burials beneath tons of rock and then the bodies savaged by dogs, birds and beasts were thrown hurriedly into the river. The Harijan Saroo was not of them—he had no afterlife, no chance of salvation....

At the Goswami Ganesh Dutt Saraswati Mandir at Uttarkashi the handbills announcing the "Sree Narayan Bali, the Ekkis Sreemad Bhagavad Gita Parayan and the Mukti Yagna (yagna for salvation) was categorical: The village pradhan should specifically mention the caste, *gotra* and name of the dead. The invocation read: *'Sarve Bhavanthu Sukhinah, Sarve Santu Niramayah, Sarve Bhadrani Pashyantu, ma Kaschit Dukh Bhag Bhavet'* (Let happiness and health be with everyone, let the eyes see only the pure, let there be no sorrow)....

At Maneri, Umed Singh Panwar stands amidst the ruins of his three storeyed-twenty-five room mansion. The idols are lined up in the open, looking for a new abode. The Bhagirathi flows along his land in quiet turbulence, swallowing a bit of life with every movement..... The house in front collapsed killing two children. Umed Singh Panwar and his family are safe and he looks around the material ruins with a new faith in his idols of stone....

The night of the earthquake his entire family had gathered in the "puja rooms" and he still does not know what made them decide to spend the night in those rooms, rather than the five bedrooms which they normally occupied.... "Perhaps it was the fact that the dog was uneasy and the cats were unusually active.... my brother's children were also here... there was nothing for us to do except chant God's name, as we saw the village crumbling...

The two third floor rooms still remain intact almost floating on barely existent walls while his house stands in ruins....

Relief vehicles rushing past, stop by to drop clothes and food. Panwar and his family accept the gifts with shocked gratitude.... for in the ruins they have found a new life....

Fig. 4.3.1

Courtesy: Hindustan Times

"What have I to live for..." the fifty-five-year-old Brajdevi face distorted with grief cries out. Both her sons were buried alive. Her five-year grandson plays with a dried pine core looking forward to life amidst the stench of death.....

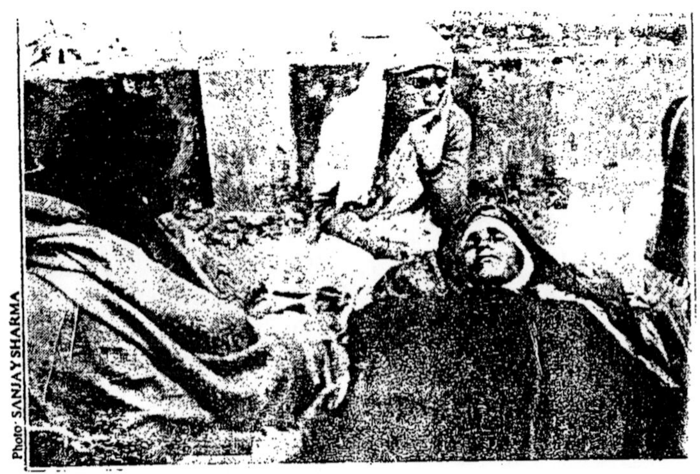

Fig. 4.3.2 *Courtesy: Hindustan Times*

A kilometre from Bhatiara, the road lies blocked. A shivering mountainside has shaken off huge boulders and the soft drum roll of death echoed in the mountainside. Over a hundred and fifty people looked towards the mountain top fascinated by fear.... half expecting another tremor and numbed by the blankness of sudden death.

There is an urgency in the "whistle" of the Border Roads Orgai.isation supervisor signalling danger as he gives the sign to "blast". A muffled sound, a huge rock breaks into a thousand fragments and the labour rushes into clear the road...

The "all clear" whistle sees the road open once again. Trucks, vans, matadors and human beings rush towards the tehsil headquarters of

Bhatiari where pack mules wait with bred patience for the trek upwards though the mountain bridle paths....

The tinge of white already appearing on the mountain tops add an urgency to the convoys carrying relief materials—any—day, any time now the snow would start falling, blocking the roads and path to the 'higher villages'.

The shepherds had started moving down long ago—the shorn sheep moving to the lower heights seeing with the contempt of the living the race against death....

"There is nothing there now... there never was..." The mountain shepherd with a lamb cradled in his arms whistled softly to his flock.

Doctors who had moved in to rescue pilgrims from Gangotri talked of the trail to Gomukh, the source of the Ganges, being completely wiped off. At Harsil, the villagers were emphatic: There were cracks in the glacier from which the Ganga starts as a trickle.... "if the glacier cracks only the Gods can help us from being swamped by floods this summer....."

The Gangotri temple was intact, though the *prangan* (courtyard) had been completely damaged.

"The Bhagirathi cannot be tamed by man," said the seventy-year-old Bula Ram, "it needed Shiva's Jhata to control it... now man tried to equate himself with God... no other river has the power of the Bhagirathi.... this dam and all this disturb the abode of the Gods...." At Harsil 8,500 feet into the mountains they do not want to disturb the Gods...

The ecologists and the environmentalists have taken the earthquake to relaunch their struggle against the Tehri project.... "The 1803 earthquake killed twenty-five per cent of the population of Garhwal, and after that we were enslaved by the King of Nepal.... 200,000 Garhwalis were sold as slaves. This earthquake is merely the precursor to our becoming slaves of construction companies and commercialism...." the well-known environmentalist then turned his best profile towards the camera....

It is not only the environmentalist or the absolute traditionalist who see in the earthquake the anger of nature against the targets of progress set by man. In Jamak, the worst affected village, a kilometre above the Manari hydro-electric project, the residents blame the project for the destruction of the village.... "They have dug the earth for the project and

Fig. 4.3.3 *Courtesy: Hindustan Times*

hollowed the mountainside, there is no earth below our village and that is why we have died...."

Yet ironically though electricity is seen as an essential ingredient of life, the cracked edifice of the Manari dam is seen as an intrusion, a monument to contractors from the plains.

At the tehsil headquarters the young Manendra Singh Rawat, a member of the District Board, was emphatic that there had been no electricity after the earthquake essentially because the local labour had been ignored and a "contractor" employed to profit from the darkness of the hills.....

Today the darkness of the hills after dusk is broken by a stream of vehicular headlights... as the convoys of relief materials attempt to bring succour to the affected....

A belaboured district administration swamped by criticism has been fighting against long odds to reach relief to the remotest corner. The Army, ITBP (Indo-Tibetan Border Police) and the BRO (Border Roads Organisation) have become folk legends for their spontaneous reaction in rushing to villages cut-off from every mode of help.... For the first three days they reached affected villages airlifting the injured, providing food and where possible, makeshift tents. Even now people wonder why they have not been used to systemise the relief efforts....

To the district administration's high profile relief effort not be given the credit both for rushing food and medical aid — almost every village was given basic foodgrains and blankets within seven days of the disaster.

It is the spontaneity of the relief effort by the voluntary organisations which have surprised the hill people — and angered those who live in villages approachable only though difficult mountain trek-paths. The villages on the roadside invited instant sympathy and truckloads of relief materials gathered hurriedly by mohalla, locality, and factory relief committees were delivered individually and haphazardly. Hurriedly improvised road-side relief stations were mobbed by the villagers — those strong enough to trek down were overburdened with relief while the affected in the hill-top villages waited for skeleted government issues.

Yet even in the succour to the families of the dead and dying there were elements who wrote their own dirty epitaph: one organisation had sent stinking, tattered blankets and quilts which were obviously garnered from the *raddi bazaar* and were sure carriers of pestilence and disease.

"The probably wanted to claim the Income Tax relief for donation." The Deputy Collector of Uttarkashi, Mr. R.P. Singh commented as he sat in the relief control-room trying to shift need from want in the clamour of the crowd outside...

It is not the quantum of relief which is the problem facing the district administration racing against time and nature... it is the shortage of the requisite type of relief materials needed. The basic needs of food and clothing have been delivered.

In perhaps one of the best examples of crisis management, an outbreak of epidemic has been avoided by massive voluntary innoculation, water purification and preventive medicine drives. A month after the earthquake, even after most of the dead bodies had been thrown in the river and animal carcasses buried in the villages immediately after the quake there has been no reports of an outbreak of cholera. "We are keeping our fingers cross," a visibly tired Dr. G.S. Rawat, the Chief Medical Officer of Uttarkashi stated while he looked at his young team...

The dichotomy of the nature of relief material was narrated by doctors who had been working round the clock: "In one village, we found villagers in the possession of capsules... distributed by voluntary agencies... no government can afford the antibiotic capsules which cost almost Rs. 18 each.... and in yet another the hoarded `relief medicine' of one villager was enough for the treatment of five others with enough left over for emergencies...."

Food, clothes and medicines would be meaningless aids for life without the primary requisite of shelter—and it is here that relief efforts are far behind what is needed. With nearly forty thousand houses uninhabitable, the flow of tents and tin sheets needed for temporary dwellings "is woefully inadequate". Some voluntary organisations have tried to do their mite which is a mere drop in the ocean. The government which has increased the number of tin-sheets to be given from four to eight is woefully behind schedule.

The villagers in the upper hills wait for the next convoy of tin sheets and shiver while Lucknow sneezes over which 'tender' to accept. The relief effort could be nullified by the "supply-cut-back syndrome" of the tin-sheets and tarpaulins.

At the tehsil headquarters village pradhans refused to lift the tin-sheets which would bend at the pressure of a finger. "You expect this to stand up in a snow-fall... we would rather be buried under a mountain slide, then beneath a tin-sheet.... "forty year Gopal Singh Rawat shouted amidst approval from the crowd. The overworked Tehsilder of Bhattari, I.S. Kumar watched silently while the "government issue" was being tested against the truth of 'shared-profit' officials and suppliers.

The problem of shelter is not only against the elements—for without it there are bound to be frozen bodies, but also against wild animals. At Tihar gaon, Madho Singh swore that a bear had quietly taken his relief supplies, including children's clothes. Reports of *baagh,* panthers carrying away depleted livestock seemed to be accepted by village officials, with a case of an attack on humans reported from Bukhi village.

The problem of shelter is intrinsic to the people of the hills as it involves the livestock which is their primary wealth. "Four tin sheets is not even enough for shelter for our cows and buffaloes... leave alone humans, and what happens if our cattle die...", Narendra Kathait was emphatic. The government has sanctioned Rs. 1250 for a cow or buffalow against the average price of Rs. 5,000 to Rs. 8,000 each. Is it sufficient to replace the livestock and thereby the prime source of existence? District officials merely shrug their shoulders with a: "Decisions are taken in Lucknow..."

In the nature relief operation the neglect of livestock could have serious repurcussion — with no fodder for the cattle, the people may well have to choose between their livestock and life..."

In the Uttarkashi region the anger of nature could well turn into anger against the government. "They want to turn the region into a reserve forest, they are releasing animals here.... the animals can kill us, but if we kill a panther or a bear we will be imprisoned.... let the government decide whether they want to rehabilitate the animals or us...." Manendra Singh Rawat was obviously voicing the sentiments of the villagers from Barsi.

Already the politicians, irrespective of his party is persona *non grata* village pradhans are careful not to mention party affiliations though the caste of the pradhan often determines the distribution of relief in the village — with the Harijan being the last beneficiary. There are however committed examples of dedication like that of the 36-year-old Mrs. Maheshwari Bhatt, the woman pradhan of Nepala village who has earned the affection of every village irrespective of age or caste.

The earthquake in its wake has also brought out some of the primeval instincts of man—the Darwinism manifest in the strongest cornering the relief irrespective of whether the weak or the aged lived or died was the normality rather than the exception. The quite confident Pahari proud of his home amidst the Devasthan, "abode of the Gods", is now as scared of life as he is of death.

"Resettle us somewhere in the area between Rishikesh and Dehradun," the sixty-five-year-old retired school teacher Sachindanand Nautiyal demanded, "for the hills have been made unsafe to live and unlivable to die in peace...."

Outside the *tehsilkot* the pack mules waited patiently. "For God's sake get the bulldozers to clear the Gangori-Dodital road otherwise the people will die...' the tehsildar shouted.

"There up there", said the young Lalita Prasad as he pointed to his village Kumalti, a kilometre above Mucheri, "a crack a feet wide has encircled our houses.... how do we live... when will we die..."

Decision takes a long time to die.

Courtesy: The Hindustan Times

4.4 WAS THE CATASTROPHE MAN-MADE?

—H.B. Singh

The Uttarkashi-Chamoli earthquake on October 20, 1991 has caused planners, environmentalists and Government agencies to take a fresh took at the fragile ecosystem of the Himalayan region. The Himalaya represent the youngest of the folding mountains in the sub-continent and its ecosystem is being nibbled and altered due to the pressures of increasing population in the region. Seismologists believe that the damage due to earthquake need not necessarily depend on the magnitude of the tremors alone, and that several other factors such as the kind of area, type of dwelling structures, soil conditions, time of occurrence and habitation play an equally important role.

Fig. 4.4.1 *Courtesy: The Hindustan Times*

It is well known that shifting cultivation, increased demand for fuel, fodder and packaging material for fruit and vegetables, excessive grazing, and developmental projects have resulted in severe problems of soil stability, water resources and survival of biological diversity. The magnitude of loss of life and property in the recent bio-mishap in the Himalayas is directly or indirectly the result of the same ongoing clash between ecology and economy in the region.

The National Forest Policy of India states that in the hills at least 60 per cent of the total land area should be under forest cover. Unfortunately, the forest cover in the Himalayas is much less than the desired percentage. An assessment made by the Forest Survey of India by interpretation of Landsat imagery indicate that only 16.71 million hectares of land in the Himalayas is under actual forest cover. This constitutes only 37 per cent of the mountain area. Of this forest cover, 7.65 million hectares (15.21 per cent of the total area) is not adequately stocked.

Most of the energy requirements of the hill people are met from the forest biomass. A regional study conducted by the FAO recorded that the Himalayan region in Asia is one of the six zones facing a serious fuel-wood crisis. The pressure on forests for fuelwood has crossed the sustainable limit in India and now it accounts for 1 million cubic meter as against the prescribed limit of 0.7 million cubic meter.

The Chipko movement under the leadership of Sunderlal Bahuguna and Chandi Prasad Bhat took its birth following large scale landslides that shook the upper ranges of the Garhwal Himalaya. The worst tragedy occurred in 1978 when the Bhagirathi unleashed a mammoth landslip in the Gairaidhar catchment area of Kanolddikhad creating two ephemeral lakes that eventually destroyed several villages. The striking feature of the movement was that the women-folk, who were the most affected by the degradation, came to embrace the trees till the forest contractors left the scene. The movement got world attention but failed to gain enough momentum. Moreover, while enough awareness has been created to present felling of trees by 'outsiders', not much has been done to safeguard trees from being butchered for local use.

The hill ecosystems are generally fragile and cannot sustain heavy biotic pressure. It is a well known fact that biotic pressure, faulty land use system and mismanagement of natural resources are the chief causes of degradation of the Himalayan ecosystem. The guiding factor in the hills should be the bearing capacity of the site and not the geographic area available for exploitation. Therefore, the population flux into the hills needs to be discouraged along with a check on industrialisation and urbanisation.

We need to develop specific guidelines for the developmental activities in the hills. Road building has often led to destruction rather than development in several areas. Detailed studies need to be carried out on the regeneration potential of the site; site productivity; effect of exploitation on the site; water quality; erosion; and regeneration. A glaring example is the extraction of timber in the high level conifer forests in the Western Himalayas which has led to regeneration problems of fir and spruce.

The livestock population in the Himalayas is too heavy and most of the cattle are unproductive. We need to reduce pressure on our grazing lands. People need to be educated about the false social status attached to the number of cattle head possessed. The cow breed can be improved through hybridisation. Pasture management and stall feeding are inevitable if the mountains are to be saved from becoming heaps of dust and boulders. A ban on open grazing would automatically bring down the number of unproductive cattle.

Faulty land-use in the Himalayas is a very complicated and multifaceted problem, compounded by increasing population. The land-use practices prevailing in the Himalayas have never taken into consideration the vital role of these mountains in the overall geo-climatology of the whole of the Indian subcontinent. It is not a simple chain of hills but harbours some of the biggest watersheds of the world. Many rivers emanate from these hills and fulfil the water needs of the vast plains below. The vegetation ranges from tropical to alpine with a great diversity of species with a very high percentage of endemics. Some of our finest wildlife areas are in these mountains or along their foothills.

Faulty land-use practices have led to overall degradation of these mountains. At present agriculture and horticulture have been extended to very steep slopes especially in the Western and Central Himalayas. Hills are ecologically suitable for perennial cover and repeated crop harvesting and tilling on steep slopes renders the soil infertile. Unfortunately more and more area is being brought under agriculture and there is even a feeling among the hill people that their region should be self-sufficient in food production. However, in the long run, only agro-forestry, based on perennial species, can be successful and productive in the hills. This system is more akin to the nature.

Horticulture, has played a great role in the economic development of the hills and therefore, has been extended even to some unsuitable sites. This needs to be controlled. The use of wood for packing horticultural produce is one of the most wasteful practices, and it should be discontinued immediately. A short-term solution can be use of fast growing species planted in the plains and then transported to hills for use in packing. An experimental effort in this direction has given encouraging results. The Haryana government is supplying eucalyptus wood to Himachal Pradesh for apple packing cases. Post harvest research in fruit and vegetable packing needs to be stengthened and some reusable plastic-like materials should be tested.

A long range solution can be found by encouraging the plantation of low volume but high income yielding plant species such as walnut, hazelnut, chestnut, almonds, etc. which do not need wood for packing. One should strongly recommend the further expansion of horticulture to the extent of only nut yielding species, to conserve the Himalayan ecosystem. Fruit production can be increased by grafting economic cultivars on wild species e.g. pear on kainth. Chilgoza has been successfully grafted on chir pine and preliminary efforts have given good results in this direction.

Tree plantation is now synonymous with environmental conservation, whereas it is the optimum land-use which is important. Any planning should take care of two things — one, what the land can sustain and secondly, what are the social needs. The best strategy would be to consider the site potential and limitations and then decide the choice of species. The failure of many of the afforestation programmes has been due to selection of wrong species from social as well as site point of view. Whatever species are selected, site improvement should be an important objective, along with increase in productivity, on a long-term basis. Agro-forestry has been recognised as the only viable long-term sustainable system with proven social, economic and ecological benefits. The selection of suitable site-specific species can be achieved by meaningful research and trials.

Courtesy: The Hindustan Times

4.5 MAJOR REASONS OF DESTRUCTION IN GARHWAL HIMALAYA

250 villages of Uttarkashi district out of 681 severely damaged by the earthquake shocks of 20th October 1991. According to local residents only earthquake shocks were responsible for such destruction in Garhwal, but it is not true, because human activities which were done to fulfil their requirements of physical happiness also contribute a lot during this natural disaster. Few of the important reasons which were responsible are being discussed hereunder.

Deforestation

Vegetation cover on the earth surface plays an important role in balancing the top soil against surface removal through root system. A large scale exploitation of forest resources is done to fulfil the ever increasing demand of agricultural land, fuel, raw material and infra-structure facilities.

Most of the Himalayan region is already prone to landslides and landslips due to geodynamic unstability and reduction in vegetal cover further aggravated the situation. That is why, during the earthquake shocks a large segment of rock and soil sliped down and played an havoc with various ecological components. This could be well explained by Fig. 4.5.1.

Dam and Road construction

In Garhwal, roads are the most rapidly expanding developmental phenomenon affecting the environment in a manner which is far away from positive one. There are several river valley projects which are either under construction or constructed. The controversial Tehri Dam is the biggest of them. For the construction of roads and dams, unscientific blasting of rock structures by using dynamite mainly results in high landslide rate due to the cracks in rock structure. According to the residents of Jamak village, they experienced shocks and shaking of village on every blast of dynamite during the construction of Maneri Dam. By this, we can say that alarming cracks which were appeared due to such activities amplified the extent of destruction.

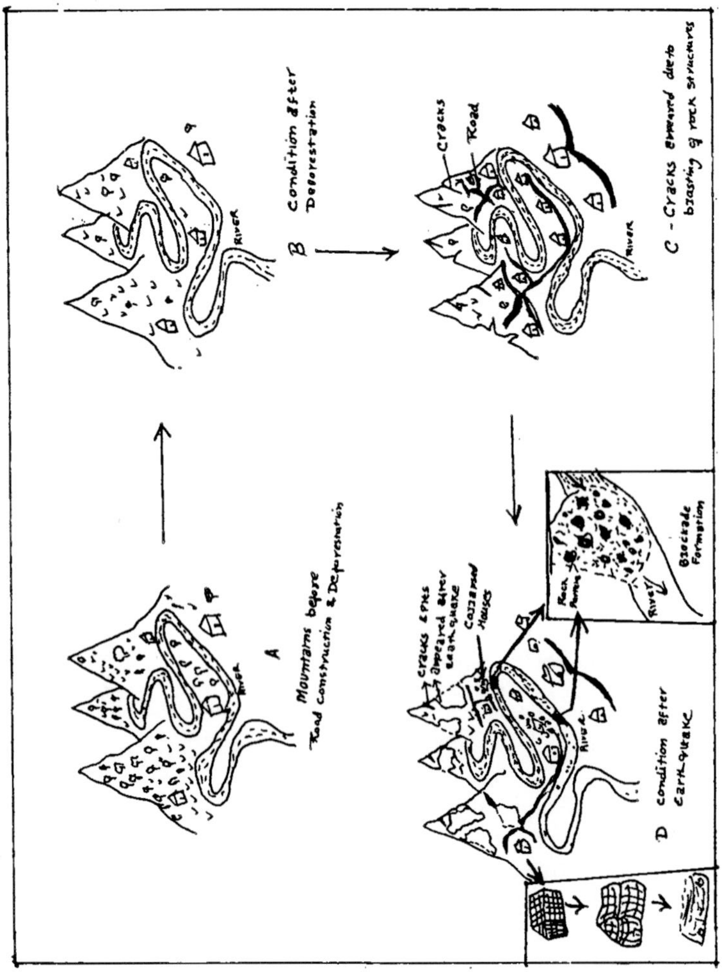

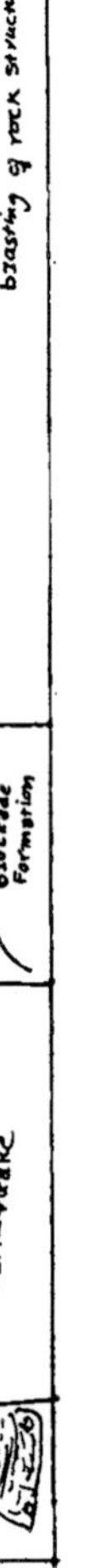

Fig.4.5.1 Earthquake and Damage

Quarrying

Unscientific quarrying of stones in this region has also caused profound adverse effect on the cyclic behaviour of the natural ecosystem and promoted its physical and biological degradation through increased rate of landslides and landslips. The large pits formed as a result of mining and quarrying are the prominent sites for slope failure during earthquake shocks.

Illy constructed houses

The number of houses has been increased manifold during the recent past due to the fast growth of tourist industry in Garhwal, which were constructed mainly by using easily and cheaply available construction material (like—silt, stone, wood, slate stone, sand etc.). Walls of house were constructed by using stone in place of brick and sand in place of cement as cementing material, without caring the strength of base and wall. On the walls roof was prepared by putting slate stone in titled position with the support of wooden structure. During the earthquake weaker walls were not in a position to hold the roof structure, due to which heavy roofs settled down and complete destruction of house took place.

4.6 LARGER DAMS IN GARHWAL HIMALAYA : A RESTROSPECTION

N. Singh, K. C. Bhatt & C.P. Juyal

INTRODUCTION

The Garhwal hills of the Central Himalaya, owing to their physiotopography, geographical location at the head of great Gangetic plains and climatic conditions are the source of a number of perennial rivers which finally emerge as the Ganga, the life line of a large area of northern India as well as the witness of the culture and civilization since the times of immemorial. The upper riperine basin of the Ganga (in the uplands of the Garhwal region) are full of immense economic potentialities, *viz.*, drinking water, irrigation facilities, household purposes, industrial uses, recreation, hydroelectricity, fishery development, tourism etc. These potentials are to be utilized with suitable technology and prudent planning for the benefit of the country.

Consequently, the Garwhal region is, at present, the theatre of many multipurpose river-valley projects involving almost all the major perennial rivers. These projects are mainly devoted to the prime cause of hydro-electricity power generation (Table 4.6.2), but these will also initiate the chain of equally beneficial activities. The development work on most of the river schemes have also led to several allied activities so as to create the required infrastructure, thus, aggravating the ecological imbalances of the region (Singh et al. 1992). The earthquake of 19-20 October 1991 was an alarming signal for the wrong deeds being committed with the nature. The earthquake has generated fresh hot debate and controversy regarding the wisdom of continuing with these large scale major projects. The main concerns appear to be the safety and security of the structures, people, climate, ecology, fauna and flora.

This article is devoted to initiate the alternative approaches resulting from the need of harnessing the hydroelectricity potentials (through large scale dams and reservoirs) and unpracticability of major projects.

PHYSIOGRAPHIC FEATURES OF GARHWAL HIMALAYA

As geo-political unit of the Central Himalaya, the Garhwal region is the western part of Uttar Pradesh hills situated between the latitudes 29°26' - 31°28' N and longitudes 77°49' - 80°60'E with a total area of about 30,090 km². It comprises five northern districts of Uttar Pradesh — Chamoli, Uttarkashi (larger and border districts), Tehri, Pauri and Dehradun. The northern region extends up to the snow clad peaks as a part of Indo-Tibetan boundary; in the east, it touches the borders of districts Pithoragarh, Almora and Nainital (Kumoun division); the southern borders are common with Bijnore district (Rohilkhand division), Hardwar and Saharanpur (Meerut division) while river Tons and Yamuna separate it from neighbouring Himachal Pradesh (Fig. 4.6.1).

On the basis of the physiographic attributes such as absolute and relative reliefs, it is divisible into three physiographic regions, 1. the Greater Himalayas with snow clad peaks with a height of over 7,000 m with well as snow ladden valleys, 2. the Lower Himalayas with high mountain ranges, hills, valleys and lake basins, and 3. the Sub Himalayan Tract or the Shivaliks including the Shivalik ranges and 'Bhabhar' (foot hills with a height of not more than 325 m). The Upper and Middle

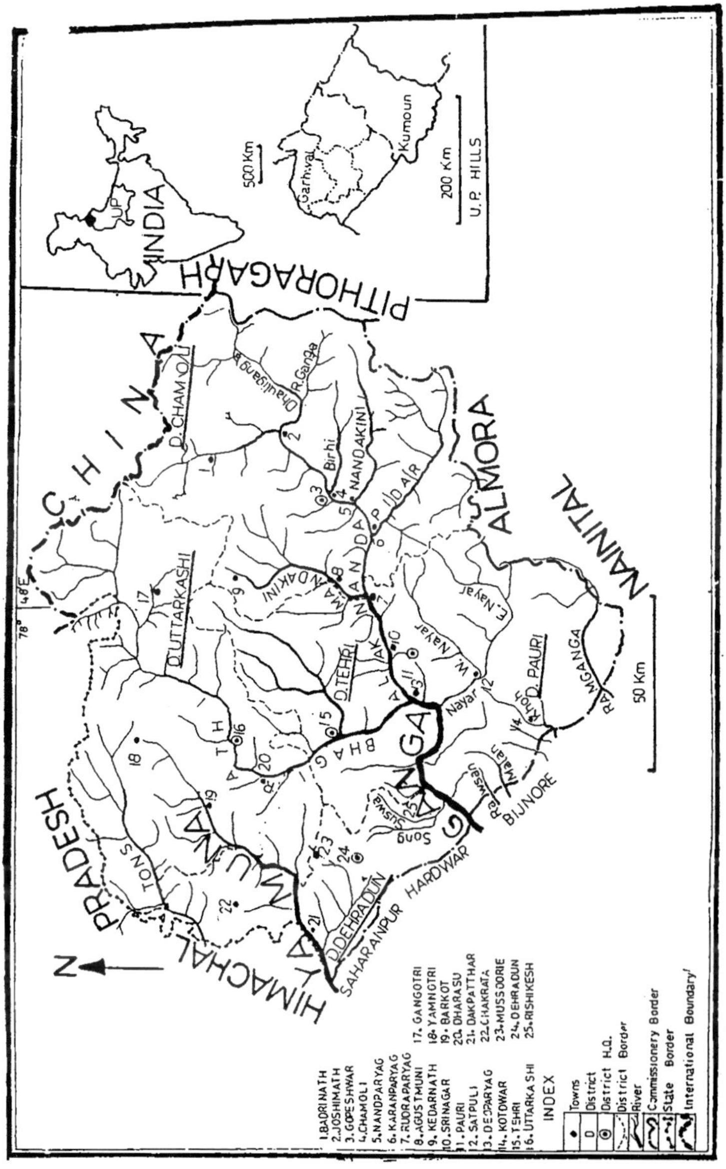

Figure 4.6.1 Drainage Pattern in Garhwal Himalaya

Himalayas in the Chamoli and Uttarkashi districts are characterized by a number perpetually snow bounded glaciers — Gangotri (30°45' - 30°55' N and 79°05' - 79°15'E; elevation about 3,900 m approximately 30 km long and 2 km wide with a system of tributary glaciers as Rakta Varna, Shwet Varna, Nilamber and Pitamber), Chaurabari (30°50' - 31°0' N and 79°0' - 79°05' E; above Kedarnath peaks; 1,400 m long and 500 m wide), Satopanth (30°41' - 30°47'N and 79°19' - 79°25' E; south west of Badrinath and Kumaling peaks) Bhagirath Kharak (30°48' - 30°50'N and 79°15' - 79°25'E; besides north and south Rithi, Yamunotri, Pinder, Juna and Nandadevi glaciers. It consists of magnificent series of glaciers garlanded peaks — i. Benderpunch (6,315m), ii. Chaukhamba (7,138m), Gangotri (6,614m), Kedarnath (6,940m), iii. Kamet (7,756 m) and iv. Nandadevi (7,817 m), Dunagiri (7,066 m), Trishul (7,210 m), Nandakot (6,861 m) etc. These four groups of peaks have been separated by the transverse gorges of the Bhagirathi, Alaknanda and Dhauliganga. Thus, the Garhwal Himalaya is the eternal home of glaciers, horned peaks, serrated crests of high ridges, creques, cascades of sparkling water coming from melting snow, torrential rapids, deep canyons, roaring streamlets, huge boulders and glistening lakes (Kharakwal 1977).

The glaciers and snow clad peaks in the Greater Himalaya (Chamoli, Tehri and Uttarkashi districts) are the sources of numerous perennially and eternally flowing rivers which ultimately emerge from Uttar Pradesh Garwhal hills as the Yamuna, the Ganga and the Ramganga along with several spring fed rivulets.

The Garhwal region is well drained by numerous rivers and rivulets (locally called as 'gad' and 'gadhera') (Fig. 4.6.1). Practically all the glaciers and valleys individually are the sources of the one or more streams major or minor dimensions. The drainage system of Garhwal region can be divided into three main components (which emerge out of southern fringes of Garhwal hills and these follow their own courses in the plains of Northern India). These are —

1. The Ganga river system : Major part of the region is drained by the Ganga except western parts of district Uttarkashi, Tehri and Dehradun. In fact, it comprises —

   A) The Bhagirathi basin — the river Bhagirathi tickles down from the Gomukh (3,900 m) in the Gangotri glaciers and western face of the Chaukhamba peaks (within the physical boundaries of district Uttarkashi). Before coming down to Uttarkashi town (1,158 m), it receives the Jarganga and Asiganga, other glacial and non-glacial fed streams. During its further course, the Bhilangna (glacial fed originating from the Khatling glaciers — 3,950 m in Tehri district) merges with it at Tehri (630 m), till further downwards at Deoparyag (472 m), it meets with equally important sister tributary of the Ganga — the Alaknanda. Throughout the entire journey, all the streams of basin pour themselves in the Bhagirathi. Significantly, the Bhagirathi and Jarganga have cut awe-inspiring gorges through tourmaline granites constituting the central axis of Himalayas. The Bhagirathi basin comprises the parts of district Tehri and Uttarkashi.

B) The Alaknanda basin — the Alaknanda originates from the Alkapuri glacier (a combined snout of the Bhagirathkharak and Satopanth glaciers, eastern face of the Chaukamba in the Chamoli district) and while taking a treacherous course through deep gorges and collecting many glacial and non-glacial fed streams, it meets with the Dhauliganga (which itself drains a large hilly basin) at Vishnuparyag (1,372 m). Afterwards, the Alaknanda flows sown through Nandparyag (914 m confluence with the Nandakini), Karanparyag (795 m, confluence with the Pinder), Rudraparyag (610 m, confluence with the Mandakini) and finally to Deoparyag (472 m) to merge with the Bhagirathi and contributes to the origin of the Ganga river system. These rivers (just mentioned tributaries of the Alaknanda) are themselves quite large and require separate description. However, the Alaknanda catchment area comprises the district Chamoli, parts of district Pauri (north west) and Tehri (north east) besides north western fringes of Kumoun division.

Evidently, the Ganga results from the confluence of the Bhagirathi (265 m) and the Alaknanda (200 m) at Deoparyag. The combined stream traverses down through Byasi (495m), Gular, Laxmanjhula, Rishikesh and Hardwar (295 m). On the way, it is fed, on eastern side, by the Nayar and other smaller spring fed tributaries and, on the western side, by Gular, Song, Suswan etc. The catchment area includes district Pauri (north eastern, central and south western parts), Tehri (eastern and south western parts) and Dehradun (eastern parts).

2. The Yamuna-Tons river system : The river Yamuna owes its origin from the Yamunotri glacier (3,300 m) lying on the south eastern slopes of Banderpunchh peak (in Uttarkashi district). It collects a

large number of streams on both sides, while proceeding through Purola, Barkot, Kalsi, Lakhwar, Dakpatthar till it reaches Paonta Sahib after covering a distance of over 150 km through deep gorges and valleys. The river Tons is the biggest tribuntary of the Yamuna; it originates from the northern slopes of Banderpunchh and flows through a valley, northwest of Yamuna, so as to joins the latter below Kalsi. It brings nearly double the volume of water of Yamuna.

3. The Ramganga river system: Of the Garhwal region, only south western parts of Pauri district comprise the Ramganga basin, otherwise, it is the drainage system of Kumoun division.

The Shivaliks form the southern fringes of Garhwal hills. While in Dehradun district, it hardly allows the passage of any stream through it, but in Pauri district, the Shivaliks are traversed by many spring-rain fed streams particularly Rawson, Malan, Khoh etc.

On the basis of source of supply, the rivers in Garhwal Himalaya belong to two categories —

a. Glacial-snow fed rivers — fed by snow melt from perpetual glaciers, e.g., Yamuna, Tons, Jarganga, Asiganga, Bhagirathi, Bhilangna, Alaknanda, Dhauliganga, Rishiganga, Nandakini, Pinder, Mandakini, Ramganga etc.

b. Non-glacial or spring-rain fed rivers — fed by the spring waters, e.g., Nayar (including Western and Eastern Nayar), Laster, Hinwal, Badiyar, Gular, Suswan, Song, Aswan, Rawson, Malan, Khoh etc.

These rivers have formed deep valleys, sometimes narrow and sometimes broad, in accordance with lithology. The dendritic pattern is most common; local radial pattern has developed around hills and peaks. Trellis drainage is to be found in the areas of thrust and faults where main streams have their subsequent affluents at right angles. The Himalayan valleys have undergone intermittent upheavals and each upheaval has caused a rejuvenation of the valleys. The evidences of recent rejuvenation are reflected by steepening of the transverse V-shaped valleys in many of the Himalayan rivers and also by rivers terraces, incised meanders and knick points in the form of waterfalls etc. Chhibber (1951) recognised the continued deepening of incised meanders of Bhagirathi between Tehri and Gangotri. Extensive river terraces are found at Srinagar-Kirtinagar and Tehri-Chamba. Three or four sets of terraces are quite obvious in the Alaknanda valley. Beautiful incised meanders occurs at Bachhelikhal and Bhaktiyana.

The climatic and meteorological conditions are affected by the location and physio-topographic configuration which collectively causes certain striking attributes to the hillstreams of Garhwal region :

1. All the main rivers have a peculiar tendency to flow for some distance in structural troughs parallel to the mountains, but sooner or later, they take on acute bend to flow in deep transverse gorges at places of hundreds meters deep.

2. Practically all these river make a steep descent in the upper reaches (at least in first ten or twenty kms) of their longitudinal profiles and afterwards the gradient does not remain so steep (Table 4.6.1).

3. The river bed is obviously uneven and comprise the rocks, boulders, stones, gravels, sand etc. The river bed may be supposed to be antagonistic to the mountain crests. It definitely provides unpredictably wide fluctuations in the hydro-median depth within the limits of short distances.

4. Steep gradient accounts for the fast water current. Therefore, steep gradient and uneven river bed coupled with the fast water velocity cause extremely high water turbulence.

5. Because of topographic location in sub temperate zone and source of water, practically all the hillstreams of Garhwal region have extremely water temperature (especially upper reaches of glacial fed rivers). However, in spite of this fact, the water never freezes even during extreme winters.

6. The differences in the nature of glacial-snow fed and non-glacial fed rivers are also evident. Recently, Nautiyal et al. (1991) statically proved water temperature and water velocity as most significant features. The coined the terms — 'torrential stenothermal' (for fast flowing glacial fed streams with narrow ranges of temperature variation without much scarcity of water during lean months and are richly supplied during summer and monsoon because of snow melting, the rivers like Yamuna, Bhagirathi, Alaknanda etc.) and 'placid eurythermal' (for relatively flowing spring fed streams with wide ranges of temperature fluctuations, not too much regular water supply especially during summer months; they are nursed by the waters from springs during summers and, therefore, the replenishment of springs in the monsoons decides the further course of water supply

Table 4.6.1

Table 4.6.1 Courses of major rivers through Garhwal region and their gradients

Sl. No.	Name of the river	Upper stretch From	Upper stretch to	Lower stretch From	Lower stretch to	Distance (approx.)	Fall (m)	Gradient (m/km)
1.	Alaknanda	Alkapuri glacier (3300 m)	Vishnuparyag (1372 m)	—	—	70	1928	27.54
		—	—	Vishnuprayag (1372 m)	Deoparyag (470 m)	130	902	6.93
2.	Nandakini	Nandadevi glacier (6000 m approx.)	Nandparyag (914 m)	—	—	47	5085	108.20
3.	Pinder	Pindari glacier (5500 m approx.)	Karanparyag (795 m)	—	—	74	4705	63.50
4.	Mandakini	Kedarnath (3700 m)	Rudraparyag (610 m)	—	—	95	3090	32.50
5.	Bhagirathi	Gomukh (3900 m)	Uttarkashi (1158 m)	—	—	120	1654	13.77
		—	—	Uttarkashi (1158 m)	Deoparyag (470 m)	145	686	4.73
6.	Bhilangna	Khatling glacier (3590 m)	Tehri (630 m)	—	—	135	2960	21.93
7.	Yamuna	Yamnotri (3300 m)	Dakpatthar (700 m)	—	—	150	2600	17.33
8.	Ganga	—	—	Deoparyag (470 m)	Hardwar (294 m)	80	175	2.19
9.	Nayar	—	—	Satpuli (680 m)	Byasghat (415 m)	30	265	8.33

through these rivers, for example, Nayar, Laster, Khoh etc.). Obviously, these are the consequences of source of water supply, gradient and amount of total water discharge which cause other differences too.

7. Based on the stratification proposed by Illies and Botosaneanu (1963), the glacial fed rivers of Garhwal region (larger rivers in particular) have been described by Sharma (1991) to constitute the rhithron zone of Ganga river system (with special reference to the Bhagirathi and Ganga, however, it is satisfactorily applicable to other hillstreams of the region). The rhithron itself comprises —

   A. Epi-rhithron : from upper altitudes down to 2,600 m. It is characterized by sharp gradient, glaciated valleys, alternate waterfalls and shallow pools, river bed of coarse large boulders. As a result of swift water flow in this zone, water directly strikes with boulders forming a white look. There is nearly complete absence of biotic components in water.

   B. Meta-rhithron : from 2,600 m down to 450 m with moderate gradient, rapid cascades — low gradient riffles at many places, river bed of small boulders, pebbles, gravels and cobble sized particles. It harbours a considerable variety of biotic components.

   C. Hypo-rhithron : from 450 m down to 300 m with a almost flattened gradient (2m/km). Main features of this zone are the dominance of pools and glides, relatively straight run of water at several places especially in lower stretches, low gradient riffles with moderate water current, river bed of sand, gravels and small sized pebbles. It is rich in different biotic components (qualitatively as well as quantitatively).

Spring fed streams comprise the potamon zone of this classification with no further sub zonation.

Many of these features have successfully been exploited for harnessing valuable potentials for the benefit of the nation.

MULTIPURPOSE RIVER VALLEY PROJECTS IN GARHWAL

Among the immense potentials embodied in the hillstreams of Garhwal region, hydroelectricity appears to be significant in view of diversifying energy resources (conventional and non-conventional), multiplying energy requirements, petroleum deficient state of India, over dependence on petroleum and petroleum products with numerous riders detrimental to the nation's prestige, glooming prospects of pollution and hazards in harnessing other energy sources and, above all, sudden changes in global scenario arising from the fast moving and uncertain ground realities in the Middle East, unpredictable series of events in the (former) USSR.

The national policy planners in India, in the early decades of independence might have foreseen such a grim scenario and they called for tapping the vast hydroelectricity resources available in the hilly regions of the country. Consequently, about 38 major or small scale river projects, mainly devoted to the cause of hydroelectricity power generation (as well as to irrigation, drinking water, tourism, fisheries, water sports etc.) are either under construction or have been proposed (Fig. 4.6.1, Table 4.6.2) with a thrust on self-reliance.

Significantly, the sites for practically all the schemes have been selected at the places where the stream passes through narrow and deep valleys. It provides better opportunities for the construction of dam or reservoir structures, features of the streams (as enumerated earlier) may be utilized in a best way so that the over all costs may considerably be reduced and the means of safely of the structures are also mobilized. Besides, other riverine, topographic, climatic etc. features are also optimally exploited for the attainment of prime and the anciallary objectives.

The perusal of data (Table 4.6.2, Fig. 4.6.1-4.6.2) reveals that these projects are spread over five districts of Garhwal region practically involving all the major rivers with the over all power generation capacity of 8776.75 MW with total capital investment of Rs. 8301.28 crores (these are the estimates in 1989 which have now multiplied many folds in terms of costs estimates as the consequences of recent currency devaluation, budgetary deficits, inflationary tendencies as well as delays in the final approval from the concerned State or Central Government agencies).

Moreover, it is also evident that of the total estimates of power generation, the maximums are spread over the district Pauri (2344.00 MW, 4 major schemes) followed by Uttarkashi (1934.00 MW, 13 major schemes), Chamoli (1763.00 MW, 9 major schemes), Dehradun (1585.75 MW 10 major schemes), and Tehri (1150.00 MW, 2 major schemes); the

Table 4.6.2

Table 4.6.2 Profile of the important river valley projects in Garhwal region*

Sl. No.	Name of the Project(s)	District(s)	River(s)	Height of dam (m)	Length of barrage (m)	Length of Tunnel (km)	Length of power channel (km)	Capacity MW	Tunnel diam. (m)	Estimated cost Crores Rs.
A.	**Projects completed :**									
1.	Yamuna HS Phase I	Dehradun	Yamuna	—	516.50	—	7.80	33.75	—	?
2.	Yamuna HS Phase I (Dhalipur PH)	"	"	—	516.50	—	5.80	51.00	—	16.83
3.	Yamuna HS Phase II	"	"	59.25	—	6.10	—	240.00	7.00	73.32
4.	Yamuna HS Phase IV (Kulhal PH)	"	"	—	288.00	—	4.00	30.00	—	14.10
5.	Yamuna HS Phase II-Part 2 (Khodri PH)	"	"	—	—	5.60	—	120.00	7.50	65.15
6.	Maneri-Bhali HS Phase I	Uttarkashi	Bhagirathi	39.00	127.00	8.631	—	90.00	4.75	79.34
7.	Garhwal-Rishikesh-Chila HS	Pauri	Ganga	—	312.00	—	14.30	144.00	—	97.76
B.	**Projects — ongoing :**									
8.	Lakhwar DP	Dehradun	Yamuna	192.00	—	—	—	300.00	—	?
9.	Byasi DP	"	"	80.00	—	2.70	—	120.00	7.00	276.42
10.	Kishau DP	"	Tons	253.00	—	—	—	600.00	—	460.00
11.	Khara HS	"	Yamuna	—	—	1.20	13.00	72.00	6.00	110.77
12.	Maneri-Bhali HS Phase II	Uttarkashi	Bhagirathi	—	81.00	16.00	—	304.00	6.00	212.66

Table 4.6.2 Continued

Sl. No.	Name of the Project(s)	District(s)	River(s)	Height of dam (m)	Length of barrage (m)	Length of Tunnel (km)	Length of power channel (km)	Capacity MW	Tunnel diam. (m)	Estimated cost Crores Rs.
13.	Pala-Maneri HS	"	"	74.00	—	12.70	—	372.00	6.00	253.07
14.	Tehri DP	Tehri	"	260.50	—	6.40	—	1000.00	11.00	1065.86**
15.	Vishnuparyag HS	Champli	Alaknanda	—	59.00	12.00	—	480.00	4.00	266.61
16.	Srinagar HS	Pauri	"	73.00	—	0.80	4.50	200.00	9.75	144.20
C.	Projects — proposed/conceived :									
17.	Arakot-Tyuni HS	Uttarkashi	Pawar	—	51.00	10.50	—	62.00	3.50	50.28
18.	Hanal-Tyuni HS	"	Tons	—	72.00	6.00	—	26.00	3.50	33.90
19.	Tyuni-Palasu HS	"	"	—	77.00	8.00	—	50.00	5.00	50.24
20.	Kuwan-Damta HS	"	Yamuna	200.00	—	12.00	—	126.00	4.00	119.00
21.	Barkot-Kuwan HS	"	"	—	30.00	17.00	—	25.00	2.50	29.82
22.	Kuthnore-Barkot HS	"	"	—	?	?	?	?	?	?
23.	Hanuman Chatti-Yamuna Chatti HS	"	"	—	26.00	6.00	—	33.00	2.50	25.25
24.	Katapatthar HS	Dehradun	"	—	196.00	—	5.95	19.00	—	27.58
25.	Rishiganga HS	Chamoli	Rishiganga	—	—	—	11.20	14.00	—	22.20
26.	Lata-Tapowan HS	"	Dhauliganga	—	—	5.00	—	89.00	2.50	40.00
27.	Markura-Lata HS	"	"	—	128.00	7.50	—	106.00	3.60	105.00
28.	Tapowan-Vishungad HS	"	"	—	55.50	11.64	—	360.00	4.80	242.20
29.	Vishungad-Pipalkoti HS	"	Alaknanda	202.00	—	—	—	340.00	—	344.00

Table 4.6.2 Continued

Sl. No.	Name of the Project(s)	District(s)	River(s)	Height of dam (m)	Length of barr-age (m)	Length of Tunnel (km)	Length of power channel (km)	Capacity MW	Tunnel diam. (m)	Estimated cost Crores Rs.
30.	Bobala-Nandparyag HS	"	"	—	102.50	9.3	—	132.00	5.60	177.40
31.	Karanparyag DP	"	"	105.00	—	—	—	160.00	—	161.53
32.	Uttyasu DP	Pauri	"	175.00	—	—	—	1000.00	—	803.00
33.	Bhagoli & Padali DP	Chamoli	Pinder	170.00	—	—	—	80.00	—	175.00
34.	Bharonghati HS (Phase I)	Uttarkashi	Bhagirathi	—	57.00	5.30	—	324.00	—	186.28
35.	Bharonghati HS (Phase II)	"	"	232.00	—	—	—	240.00	—	959.00
36.	Lohari Nag HS	"	"	—	67.50	13.60	—	282.00	4.30	177.26
37.	Koteshwar DP	Tehri	"	87.50	—	—	—	150.00	—	250.25
38.	Kotli DP	Pauri	Ganga	210.00	—	—	—	1000.00	—	1186.00
								8776.75$		8301.29$$

* Source of data, U.P. Irrigation Office, Srinagar Garhwal 1988 and Singh et al. 1991.

** In July 1991, it was Rs. 3800 crores.

$ It excludes No. 22

$$ It excludes No. 1 and 22. In view of the recent rupee devaluation, inflation and delay, the estimated will definitely increase many-fold.

Abbreviations : HS = hydroelectricity scheme, DP = dam project, — = not relevant, ? = not available.

estimated cost wise allocation stands maximum in district Pauri (Rs. 2234.34 crores) followed by district Uttarkashi (Rs. 2176.10 crores), Chamoli (1533.34 crores), Tehri (Rs. 1316.11 crores) and Dehradun (Rs. 1044.17 crores) (Fig. 4.6.3A, Table 4.6.3). The riverwise Bhagirathi seems to easily be tamed offering suitable conditions for harnessing the hydroelectricity power (2762.00 MW, Rs. 3183.72 crores) being followed by Alaknanda (2312.00 MW, Rs. 1896.74 crores) Ganga (1144 MW, Rs. 1283.76 crores), Yamuna (1043.75 MW, Rs. 758.24 crores). Others are also significant from this point of view. (Fig. 4.6.3B, Table 4.6.4).

Table 4.6.3 District-wise details of the hydroelectricity potential and proposed river projects*

Sl. No.	Name of the district	River wise break up projects*$	Number of projects	Total capacity (MW)	Estimated cost Rs. (crores)
1.	Uttarkashi	(B(6), Pa(1), T(2) Y(4)	13	1934.00**	2176.10**
2.	Chamoli	D(3), R(1), Pi(1) A(4)	9	1763.00	1533.94
3.	Tehri	B(2)	2	1150.00	1316.11
4.	Pauri	A(2), G(2)	4	2344.00	2230.96
5.	Dehradun	T(1), Y(9)	10	1585.75***	1044.17***
	Total	B(8), Pa(1), T(3) R(1), D(3), Pi(1) A(6), Y(13), G(2)	38	8776.75	8301.28

* Source of data, U.P. Irrigation Office, Srinagar Garhwal, 1989.

** excluding the figures of Kuthnore-Barkot Hydel Scheme & Yamuna Hydel Scheme Phase I

*** excluding the figures of Yamuna Hydel Scheme Phase I (Dhakrani Power House)

*$ Abbreviations : A - Alaknanda, B - Bhagirathi, D - Dhauliganga, G - Ganga, Pa - Pawar, Pi - Pinder, R - Rishiganga, T - Tons, Y - Yamuna.

The data reflects only the fraction of total existing potentials which may further be tapped with increased efficiency, rational planning, minimal ecological transformation etc. No proper estimates are available as far as the details and other information (mainly concerned with hydroelectricity power generation) regarding other rivers are concerned (Asiganga, Jarganga, Bhilangna, Nandakini, Birahi, Mandakini etc. among glacial fed streams; Nayar — including Eastern and Western tributaries, Suswan, Malan, Khoh, Hinwal, Song, Aswan, Gular, Badiyar, Laster etc. among spring fed rivers) which are fully pregnant with ample hydroelectricity potentials. It requires in depth research work to know their status and other relevant data for the purpose.

SUBSEQUENT ADVERSE TRANSFORMATIONS

The degradation in the eco-topography of Garhwal region is a part of global deterioration and a major portion of such as adverse transformation results from multipurpose river valley projects and other allied development activities. The pace of degrading transformations in this geologically sensitive zone is especially faster during last few decades because a number of transforming agents have either been freshly introduced or disturbance to the behaviours of pre existing factors. Some of these are-

A. Creation of transport facilities (road and bridge construction). Due to increased human encroachment in the hilly areas in the form of development activities, river valley schemes, tourism etc. the construction of new roads and bridges, widening of older ones to facilitate the two way traffic and efficient transport of heavy machinery and man power needed for various multipurpose river valley schemes have multiplied many-fold. It is because the roads are the only means of transport in the want of railways and airways. Roads are also significant for promoting tourism. But the menace of their construction, repairing and widening is highly objectionable and technology used for the purpose is not fully suitable to the local requirements. To accomplish this, huge amounts of fragile rock are cut/destroyed/ blown up by dynamiting and ultimately thrown in to the nearest valley or river. Such large down falling rocky lumps not only weaken the already fragile slopes but also wipe out the sparse vegetation and increase the turbidity in the rivers. The explosion due to the dynamiting weakens the age-old strength of firm rocks. All these aggravate the problems and enhance the pace of landslides and landslips, soil erosion, slope unstability which prove to be detrimental to the cause of river valley projects themselves. Presently, ongoing widening programme of Rishikesh-Badrinath road has increased the pace of landslides and road blockade at several places during last two years.

Table 4.6.4 River wise details of the hydroelectricity potential and proposed river projects*

Sl. No.	Name of the river	District wise break up of	Number of projects	Total capacity (MW)	Estimated cost Rs. (crores)
1.	Yamuna	DD(9), UK(4)	13	1043.75 **	639.24 ***
2.	Tons	DD(1), UK(2)	1	802.00	553.14
3.	Pawar	UK(1)	1	62.00	50.28
4.	Rishiganga	CH(1)	1	14.00	22.20
5.	Dhauliganga	CH(3)	3	557.00	387.20
6.	Pinder	CH(1)	1	80.00	175.00
7.	Bhagirathi	UK(6), T(2)	8	2762.00	3183.25
8.	Alaknanda	CH(4), P(2)	6	2312.00	1896.74
9.	Ganga	P(2)	2	1144.00	1283.76
	Total	DD(10), UK(13), CH(9), T(2), P(4)	38	8776.75	8300.81

* Source of data, U.P. Irrigation Office, Śrinagar Garhwal, 1989.

** excluding figures of Kuthnore-Barkot Hydel Scheme

*** excluding figures of Yamuna Hydel Scheme Phase I (Dhakrani Power House) and Kuthnore-Barkot Hydel Scheme

CH - Chamoli, DD - Dehradun, P - Pauri, T - Tehri, UK - Uttarkashi.

The taming of hillstreams of Garhwal region (geographically smaller and geologically sensitive area) have caused large scale transformations in relatively short duration. The development activities are seriously affecting the Garhwal hills' environment including hillstream ecology. For example, the Maneri-Bhali Project : a three phased hydel project with annual power generation capacity of 94.00 MW on completion of first phase, the reservoir at Maneri will have a capacity of 1294.50 MT, height and length of the 'concrete gravity dam' are 39 x 127 m; presently, only first phase (at Uttarkashi) is commissioned and others — second and third at Dharasu are still under construction. But the dam has already deteriorated irreparably the Bhagirathi ecology.

During dynamiting, tunnelling in the construction of dams, reservoirs, roads, bridges, residential complexes and site development, the rocky materials of the dimensions of dust particles to huge boulders continue to be added into the rivers. According to Singh (1987), in the process of Tehri Dam construction, 38.42 lacs cubic meters of total rocky materials were thrown into Bhagirathi and 18.75 tons of explosives were used for blasting by the end of July 1981. As a result, of over all water quality of Bhagirathi river was observed to be deteriorated. The most drastically changed parameters of water (temperature, turbidity, velocity, dissolved oxygen, free carbon dioxide) downstream the dam site are compared with that of upstream the dam site (at the end of 1981). Since then, such amount must have multiplied till today.

B. Modern tourism: Garhwal Himalayas harbours innumerable natural spots of magnificent scenic beauty — Uttarkashi, Mussorie, Panchkedar, Panchbadri, Panchparyag, Gangotri, Yamnotri besides many other religious shrines scattered throughout the Garhwal area. These are attracting an ever increasing flow of traditional as well as modern tourists. Moreover, many more sites are yet to be brought on the tourist map and developed. The multipurpose river valley projects will certainly boost up the tourism as a result of developing water sports and winter sports at many places.

It will cumulatively aggravate the deteriorating trends which are already visible even before the beginning of modern tourism. It appears that the recession of Gangotri glacier (observed during 1937-1986) resulted from the increased human activities in the form of unplanned construction of housing complex, deforestation, increased tourist inflow among other factors.

LARGE DAMS AND GEOLOGICAL SENSITIVITY

The majority of river valley projects in the Garhwal region have been proposed in the valleys of the Bhagirathi, Alaknanda, Yamuna and are very close to the epicenture of recent earthquake (Agoda, situated north of Uttarkashi town). The devastating experience of earthquake on 19-20 October 1991 (with the magnitude of 6.1 on Ritcher scale but according to California based American seismologists, it was of the magnitude of 7.1) occurred as a result of geological events originating 15 km below earth surface. The epicentre was at Agoda (initially it was reported at Almora) (Fig. 4.6.4). The tremors caused severe damage — complete collapse of houses, death of over 1,000 persons along with innumerable loss of life and property, in thousand of villages within the matter of few seconds. Broadly, the pattern of damage, in the far flung areas, may be described as

EARTHQUAKE AFFECTED AREA
H.P.
NEPAL
MCT
MBF
31° N
30° N
78° E
79° E
80° E
ALMORA
NAINITAL
PITHORAGARH
TEHRI
PAURI
DEHRADUN
HARDWAR
INDIA
UP
GARHWAL
500 Km
0
34
68
85 Km
Symbols
International boundary
State
Commissionery
District borders
District HQ
Rivers
Earthquake tremors

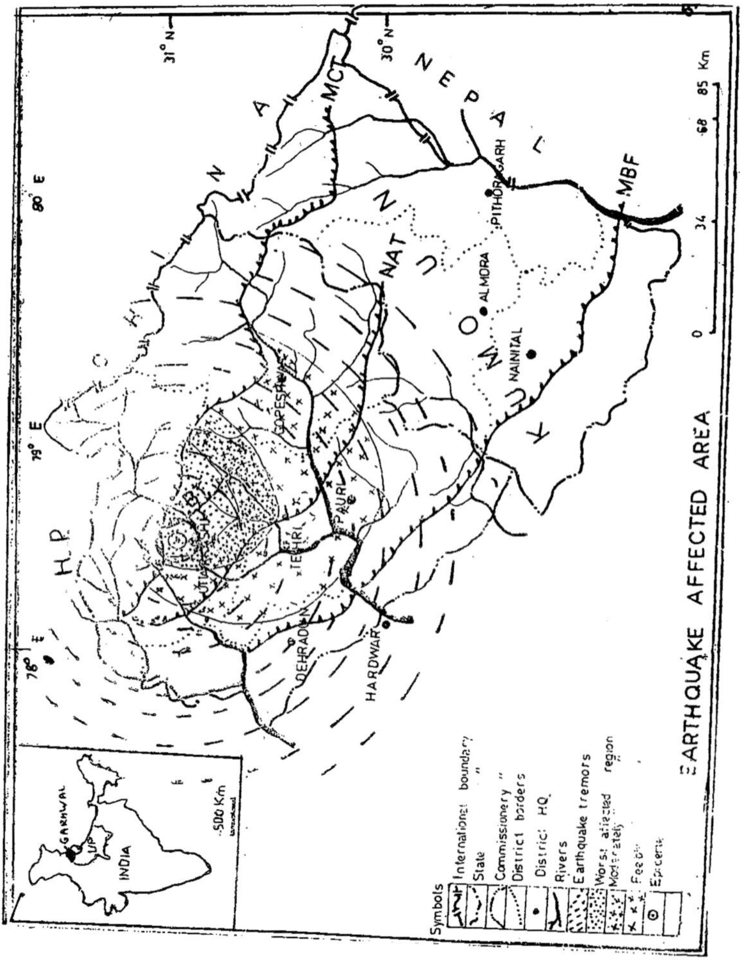

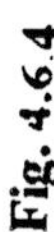

Fig. 4.6.4

A. Zone of maximum destruction : houses collapsed, triggering of landslips basically confined to district Uttarkashi.

B. Zone of moderate destruction : small scale landslips, cracks on houses, spread over district Uttarkashi, Chamoli and Tehri.

C. Zone of feeble intensity : experienced shocks and small cracks developed spread over a larger area embracing district Uttarkashi, Chamoli, Tehri, Pauri and Dehradun besides a larger region in Kumoun division and plains of Uttar Pradesh and several neighbouring states.

Earth scientists (Waldiya 1991) have identified six geological belts of this loss in terms of life and property distributed over district Uttarkashi, Chamoli, Tehri, Pauri and Dehradun.

First Belt : just south of the 'Main Central Thrust' (MCT) from Bankoli-Agoda to Buddhakedar-Thatikuthur passing through Bekelkhal-Adata (in Balganga valley), Ghuttu (in Bhilangna valley) and Triyuginarayan (in Mandakini valley). Here, high mountains had been brought down.

Second Belt : rural areas in the north of MCT and south of Vakut thrust. In the zone of MCT, it included many villages like Gorsali, Kumalti, Raithal, Pala, Morari, Tiyara, Bhukkusain (in Bhagirathi valley), Gangi (in Bhilangna valley) and Gaurikund (in Mandakini valley). These villages are located on the slides due to torrential rapids and, therefore, were the scene of large scale landslips.

Third Belt: included the rural areas surrounded by Uttarkashi thrust from Nakuri (on the banks of Bhagirathi) to Devsari-Veena through Uttarkashi, Gangori, Gawana, Maneri and Jamak. These are situated on the terraces. It appears that the changes along the MCT have been conducted on Uttarkashi thrust.

Fourth Belt: concerned with the Dunda Thrust which ranges merges with Bhatwari Thrust (from Dunga-Thati in the northwest to Silyara in the Balganga valley). It included villages like Uchhana, Gorty, Budna, Pokha, Semalta, Chone, Thela etc..(in the valleys of Nelchami and Hela). The structures in these villages had turned into rubbles.

Fifth Belt: concerned with Berinag Thrust.

Sixth Belt: related to Srinagar Thrust.

There was reported relatively scattered minor damage in the form of ordinary cracks in the house walls.

Earth scientists have explained the earthquake on the basis of 'Plate Tectonics' model of earth. Accordingly, the large middle point of trident shaped structure of Arawali continues to thrust in the Himalayas (just facing Garhwal region) resulting in the continuous underground geological changes since last 700-800 million years. Scientists are still apprehensive of more such earthquakes of the magnitude of over 1,000 times more than that occurred on 19-20 October, 1991. Such an assumption is based on the fact that Tehri-Uttarkashi structures in the earthquake prone zone directly faces the Arawali.

Interestingly, many of the river valley projects in this region (about 20 out of 38 major schemes) have been planned in the earthquake prone zone (Bhagirathi, Alaknanda and Yamuna basins in Uttarkashi, Chamoli and Tehri districts, Figs. 4.6.2, 4.6.4). For example, in case of Tehri Dam, Srinagar Thrust passes through the proposed reservoir and Nayar Thrust passes through little of dam site near Deoparyag. Both of these thrust zones have been observed to be geologically active. Predictably, the activation of Srinagar Thrust will cause the landslides on the sides of the reservoirs resulting in rapid siltation while the geological activities on the Nayar Thrust will seriously damage the dam structures. The Bhagirathi course between Dharasu and Tehri presents numerous surface features telling the story of underground changes along Srinagar Thrust. Similarly, in case of Maneri-Bhali Dam Project, considerably long tunnels are being constructed which are either very close to or directly passes through the active thrusts. The safety of tunnel structures will remain seriously doubtful. However, there was reported negligible damage to the structures of Tehri Dam or Maneri-Bhali project. But on the other hand, geologists claim that the 19-20 October 1991 earthquake is not categorized as a severe one and more quakes of much more magnitudes and severity will not be ruled out in future. Moreover, there is also the 'Main Boundary Fault' (MBF) extending from one corner to another of Garhwal region bordering Mussoorie-Pauri mountain belt and we have enough geological evidences showing that this best was frequently raised and emerged because of underground events during last 10000-12000 yrs.

Apparently, the safety and security of these projects and structure there in the zone of MCT, Srinagar Thrust, Nayar Thrust and MBF will always remain under question mark. The underground events in the earthquake prone zone are in continuum, thus, Garhwal region is truly described as 'geologically fragile area'. The ongoing development

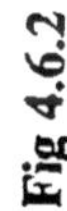

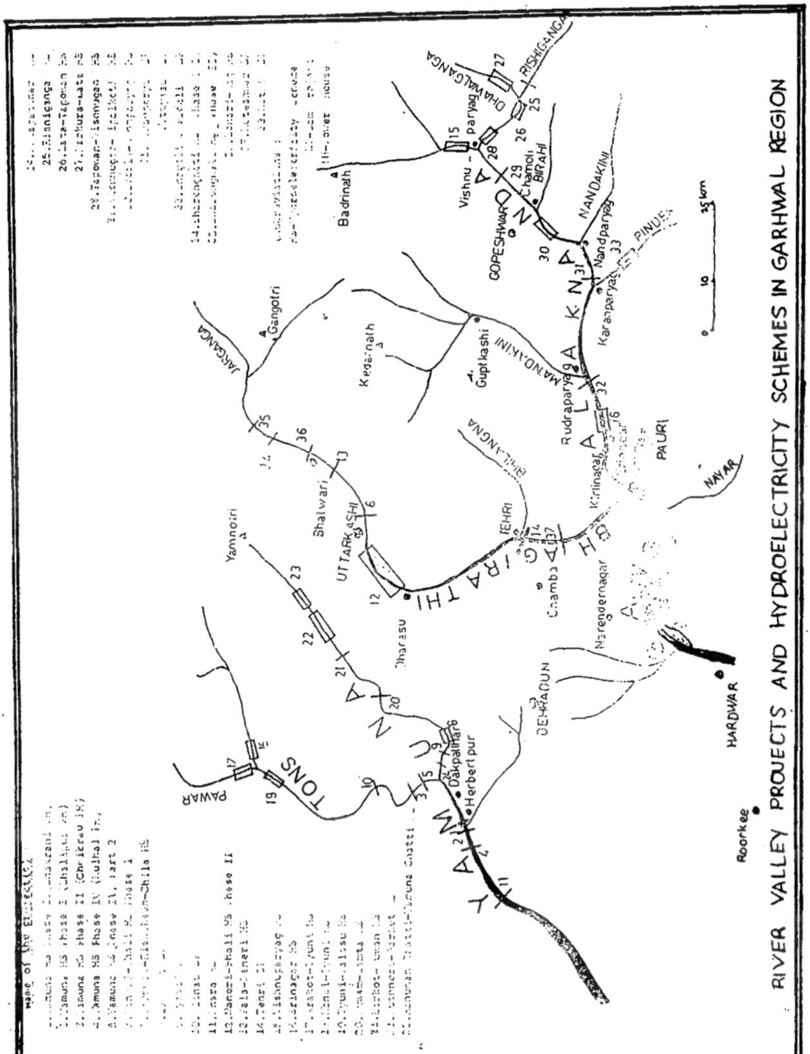
RIVER VALLEY PROJECTS AND HYDROELECTRICITY SCHEMES IN GARHWAL REGION
Badrinath
Gangotri
Kedarnath
Yamnotri
Gopeshwar
Gupt kashi
Uttarkashi
Tehri
Pauri
Rudraparyag
Karanparyag
Nandparyag
Kirtinagar
Chamba
Dharasu
Narendernagar
Dehradun
Hardwar
Roorkee
Herbertpur
Dakpatthar
Chamoli
Vishnu - Paryag
YAMUNA
TONS
PAWAR
BHAGIRATHI
ALAKNANDA
MANDAKINI
NANDAKINI
BIRAHI
DHAULIGANGA
RISHIGANGA
JARGANGA
BHILANGNA
NAYAR
PINDER
Bhalwari
0 10 35km

Fig 4.6.2

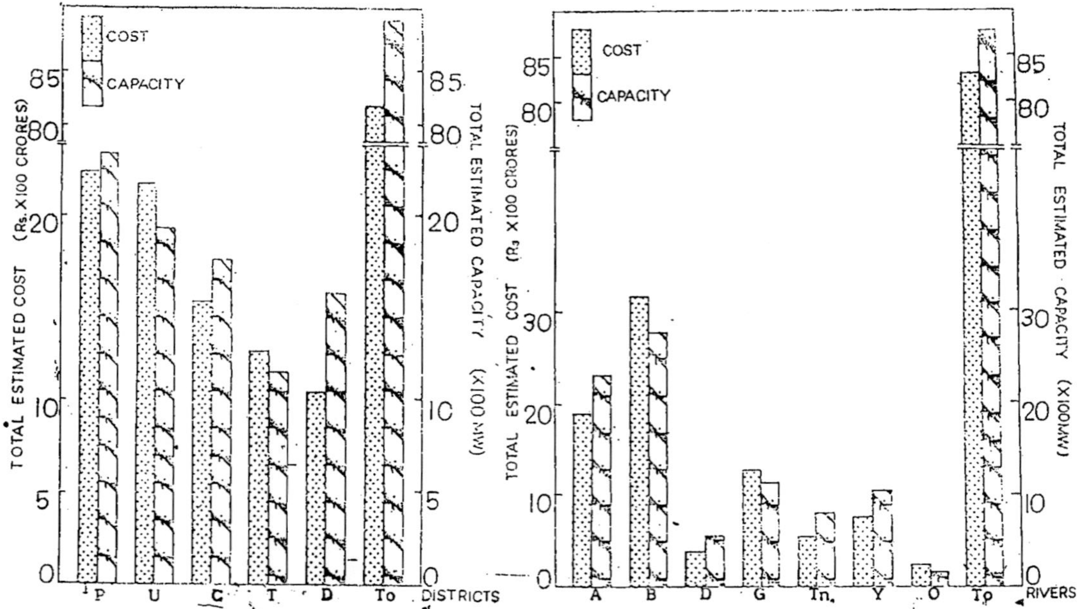

- District wise estimates of costs and hydroelectricity power capacity (P-Pauri, U-Uttarkashi, C-Chamoli, T-Tehri, D-Dehradun and To-Total)
- River wise estimates of costs and hydroelectricity power capacity (A-Alaknanda, B-Bhagirathi, D-Dhauliganga, G-Ganga, Tn-Tons, Y-Yamuna, O-others including Pawar Pinder & Rishiganga)

Fig 4.6.3A & 4.6.3b

activities have already caused an immense quantum of environmental degradation; the earthquake of 19-20 October 1991 has provided an opportunity of rethinking on the wisdom of continuing with these large scale river valley projects from every points of view.

ALTERNATIVE STRATEGIES

However, the issue of harnessing the hydroelectricity, irrigation and other potentials cannot be left unresolved and tapping of these potentials is essential pre-requisite if the national needs are given paramount importance. For this, the following points may be taken into consideration when dealing with these ticklish issue.

1. Keeping in view the physio-topography, geology, climate, ecology, safety etc., the smaller dams and small scale schemes in large numbers at short distances dotted over a larger area even in the remote places (in place of few large projects) should be emphasized.

2. Smaller dam and projects may be connected with the power and irrigation grids respectively. Such a grid system will also ensure the use of the potentials available in the smaller rivers contrary to the larger projects which always requires relatively major rivers. Moreover, failure of one small node of the grid (in the form of smaller scheme) will not hamper the over all functioning of the entire grid while little snag at one major project (in case of presently proposed system) may creste the serious problems affecting the entire grid.

3. There should be fresh survey of economic potentials of all the rivers with the clear emphasis on smaller schemes. Representatives of localities or eminent people be involved in the Rehabilitation and Replacement (R&R) programme or even informed about R & R plans.

4. The technology currently used for construction of dams, reservoirs, roads, bridges, housing complexes should further be assessed and modified to the tune of the requirement of the hilly terrain and circumstances prevailing here. It needs immediate fresh look or substitution with suitable technology favouring little or no transformations in terms of ecology, climate, topography, fauna and flora, society of the region. The construction of larger structures, whatsoever it is, needs to be discouraged.

5. Through electronic and print media, intellectuals and other pressure groups of the society should be mobilized to educate the people,

and public opinion and government agencies in the larger and long-term interest of society.

6. Lacunae of contacts and communication gap between the concerned State and Central Government should be smoothened to avoid the problems.

SUMMARY

Emanating from the perpetually snow-bounded glaciers and glacier-garlanded peaks, district Uttarkashi, Chamoli and Tehri (in the U.P. hills) are the sources of the origin of perennial and perpetually flowing rivers. These hill-streams, as a part of the upper basin of Ganga river, embody the immense economic, commercial potentials. The hydroelectricity power generation is one of the innumerable potentials. Currently, hydroelectricity potentials of these rivers is in the initial phases of exploitation (38 major schemes spread over five districts of Garhwal region along the rivers Bhagirathi, Alaknanda, Rishiganga, Dhauliganga, Pinder, Yamuna, Tons, Ganga etc.). But even during initial stages, the development activities on these schemes and subsidiary factors leading to the development of infrastructure and basic facilities) have played havoc with the health of ecology of area. The consequences are not difficult to observe and foresee in the form of deforestation, hillslides-landslips, siltation in the streams, flash floods, recurrent drought, erratic monsoons, depletion of fauna and flora (including aquatic biota) etc. Therefore, continuance with such a scenario is not prudent at all. The recent earthquake on 19-20 October 1991 has consolidated the warning of previous days and signalled the new dangers ahead including disastrous consequences. It has also offered new approaches in the direction of environmentally sustainable and geological frienly schemes. The smaller projects and small scale schemes are certainly better alternatives, a successful experience in China and Switzerland.

ACKNOWLEDGEMENT

NS and KCB thankfully acknowledge the financial assistance provided by the Ministry of Environment & Forests, Government of India, New Delhi (No. 35/20/90.RE).

REFERENCES

Chhibber, H.L. 1951 : A preliminary account of the river terraces of the Yamuna and Tons nadi in the Doon Valley, Dehradun district and intermittent uplift of the Himalayas during the recent period. Natl. Geogr. Soc. India (Banaras), 15 : 16-24.

Illies J. and Botosaneanu, L. 1963 : Problems et methodes de la classification de la zonation ecologique des caux courantes considerees surtout dee point de vue fainistique. Mitt. Int. Verein Theor. Angeno, Limnol., 12 : 1-57.

Kharakwal, S.C. 1977 : Geographic personality of Garhwal. The Himalaya, 1 : 3-13.

Nautiyal, P. Agarwal, N.K. & Singh, H.R. 1991 : Certain abiotic components of the fluvial systems and their interrelationships as evidenced by statistical evaluation, in "Advances in Limnology" (H.R. Singh ed.), pp. 181-186, Narendra Publishing House, Delhi (1991).

Sharma, R.C. 1991 : Rhithronology of Bhagirathi, Garhwal Himalaya (India), in "Ecology of the Mountain Waters" (S.D. Bhatt & R.K. Pande eds.), pp. 125-137, Ashish Publishing House, New Delhi (1991).

Singh, H.R. 1987 : Environmental Problems in Garhwal Hills, in "Science, Development & Environment" (V.P. Agarwal ed.), pp. 37-45, Society of Biosciences, Muzaffarnagar (India) (1987).

Singh, N., Bhatt, K.C. & Agarwal, N.K. 1992 : Developmental Strategies vis-a-vis ichthyofauna and environmental imbalances in Himalayas: a case study of Alaknanda catchment area, in "Aquatic Environment" (A. Gautam ed.) Ashish Publishing House, New Delhi (in press).

Waldiya, K.S. 1991 : Virat Yojnayan : Vipul Ashankayan, Bhukampgrasht Himalaya Aur Garhwal, Amar Ujala (Hindi daily), page 4, col. 3-6.

4.7 EARTHQUAKES AND TEHRI DAM

TEHRI DAM PROJECT

The most controversial Tehri Dam Project consists of 260.5 meters high earth and rock fill dam, is under construction near the confluence of the Bhagirathi and Bhilangana river, just 1.5 km downstream of the Tehri town (Uttar Pradesh). If it comes into existence will be the fifth highest dam in the world and will harness the water of two most important snow-fed rivers of Garhwal Himalaya. It will impound 3.22 million cubic meters of water. The reservoir will extend upto 45 km in the Bhagirathi valley and 25 km in the Bhilangana valley with a water spread area of 42.5 km^2. The Garhwal with an area of about 20,000 km^2 is inhabited by nearly 3 million people, out of which 30-40% are living in the valley of project. By which, about 85,600 persons will be displaced, as Tehri town and 23 villages in its vicinity will submerge and 72 villages will be partially submerged. 5,200 hectares of land will be lost to the reservoir, out of which 1600 hectares is cultivated land.

CONTROVERSIAL HISTORY OF THE DAM

The controversy on dam has begun since its very inception. The project was conceived as lastly as 1949, and on the basis of sub-surface investigation done by GSI, the site was considered suitable. In the year 1963 a detailed investigation was made, and in 1965 the site was finally confirmed. By 1967, the site was visited and investigated by several Indian and foreign experts. In their opinion the project site and nearby area is geologically vulnerable, the hill slopes which would constitute the reservoir rim area are unstable, and the seismic danger at the site. According to the project report, finalised in 1969, approximately 20 m wide fault exists along the dam site, treatment of which shall be a costly affair. It also revealed heavy cracks and large number of fault zones in the exterior zones of the slopes of the river gorge.

WORKING GROUP'S REPORT

Due to various conflicts the then P.M. directed to Department of Science and Technology to form a "Working Group" for the Assessment of the Environmental Impacts of Tehri Dam. This was constituted in December, 1979 and headed by Mr. S.K. Seth, Mr. S.K. Roy became new Chairman of the W.G. in February 1980, when an interim report was submitted in May 1980, then the W.G. realised that enough scientific data was not available on controversial aspects, particularly geological and seismic, for reaching to definite conclusions.

The W.G. therefore considered that the data so far collected are inadequate and more investigation on the below stated aspects is needed:

(a) The slope stability of the area and catchment area of the dam.

(b) Impact of Mahr fault (exposed 4 km downstream of the dam and inferred to lie at 7.5 km downstream of the dam and below the dam site).

(c) Gravity linement and tectonics.

(d) Continuous monitoring of faults lying within a radius of 100 km of the Tehri Dam, at numerous locations.

(e) Preparation of a detailed geomorphological map of the catchment area, and few others like study of landslides, landslips, soil erosion etc.

Work on the dam began in 1977, because the W.G. did not have sufficient data, therefore, he was not in a position to stop the work. In 1985, the W.G. met to finalize its report. The Chairman, S.K. Roy expressed his opposition to the T.D.P. and the final report was submitted on October 28th, 1986. This included a letter of dissent as the Engineers from Roorkee University and the Chief Engineer of the T.D.P. had, in their report, refused to support his comment (stand) that the dam was so risky. Mr. Roy also wrote a letter to the Secretary of Department of Forest and Environment (Mr. T.N. Sheshan) on August 28th, 1986, in which he has made many significant points. According to the comment 16(1) "The Tehri Dam site is not suitable for a 260.5 m high dam."

In the October of 1986, when Final Report of W.G. sent before a committee of Secretaries, it was decided that the CWC (Central Water Commission) should convene a meeting to give its opinion on Tehri Dam, specially on the issue of seismicity. According to CWC Committee (a) no major geological faults existed at the dam site, (b) adequate studies had been made to arrive at a safe and economical engineering structure to combat the risk of seismicity and (c) adequate safety factors were incorporated in the design and construction of the dam appurtenant structure.

It was March of 1987 when the Ministry of Forest and Environment observed in Raiya Sabha that the W.G. had failed to give recommendation either for its approval or disapproval. Therefore, the decision was left to the Government. The Government decided to go ahead with the I.D.P.

SOVIET EXPERT'S VIEW

In November 1986, an inter-ministerial Indo-Soviet agreement was signed, whereby Soviet expertise was made available for the construction of Tehri Dam. During 1987, various Soviet experts visited the dam site and in October 1987, the preliminary investigation was completed. According to Alexander Fink, Soviet Chief Engineer of Tehri Dam, the high seismicity of the Tehri area was not considered adequately in the Indian design, which did not make an adequate allowance for the reliability and safety of the structure. In the view of Soviety experts, seismicity at dam site to be nine-point on MM scale, while the design of the dam had provided for tremors upto M:8 only. Mr. A. Find also pointed out that the question of shore remarking and reinforcement had not been studied by the Indian Engineers.

With suitable changes, however, the Soviet Experts are confident that the Tehri Dam will be viable and will make a positive contribution to the process of the development of the region.

TEHRI BAND VIRODHI SANGHARSH SAMITI (TBVSS) ROLE

The local people have been opposing the Dam from its very inception. But the mass opposition March in 1973 when local unit of C.P.I. organised an agitation for opposing the Tehri Dam. After the Administrative approval of U.P. Govt. in 1976, thirty-five Gram Sabhas in Tehri district passed a joint resolution to oppose Tehri Dam. Without caring the suggestion of Zila Parishad to drop the project, U.P. Govt. decided to go ahead. Thus, the TBVSS came into the existence and in January, 1978, the first resolution for opposing the Tehri Dam was passed unanimously. Despite this, in March 10, 1978, contracts for the construction of the Dam have been awarded by U.P. Govt.

After the Kandolia Gad blockade on the Bhagirathi in August, 1978, a delegation of TBVSS met the then Prime Minister and suggested an alternative site for the Dam. Apparently, no action was taken on this suggestion. In August, 1980, TBVSS President Mr. V.D. Saklani made a request to Planning Commission for renewing the project but the Planning Commission responded by saying that the W.G. would look into all aspects of the matter. The TBVSS also suggested an alternative scheme for generating power by using the tunnel-cum-channel method. When no action was taken, then TBVSS filled a writ petition in the Supreme Court on November 20, 1986, challenging the soundness of the Plan to construct the Dam at Tehri. Still the opposition of Tehri Dam is on, but the Govt. decided to complete the Dam, which is now being completed by Tehri Hydro Development Corporation (THDC).

EARTHQUAKE AND THE DAM

To construct a Dam which can withstand the earthquake, is a technical matter, and a very serious controversy has been raising on Tehri Dam for several years, because the Tehri area is highly seismic. According to G.S.I. "The proposed Project site falls between isosemics of Kangra Earthquake (1905)". Dr. V.K. Gaur in his report submitted to W.G. states that "the Tehri Dam site happens to lie in a region of high seismic potential, firstly because over-thrusting of the entire Lesser Himalayan Belt southward along the Main Boundary Fault is the most dominant mode of continental convergence in the Himalaya today capable of producing a major earthquake every 300 years or so", and "secondly, because it belongs to a segment of the plate boundry which appears to be a seismic gap east of the meisoseismal zone of the 1905 Kangra Earthquake, which has not been ruptured by a major earthquake for a long time". He also says that "The Tehri region under where no major earthquake has occurred since 1928 (M 7.6) may thus have large residual strains. which have accumulated during the period". And, "since, earthquake of a magnitude less than 8.0 do not relax sufficient strain, the probability of a major earthquake whose rupture zone may traverse the Dam site is high...."

If the site is highly seismic, then the question of its stability arises. About its safety and stability or Mikhaliev, Geophysist at the Project had categorically stated that there would be no danger to the Dam from earthquake even of the strongest intensity of mine in the Richter scale, it has been designed in a manner that it can withstand these shocks. According to Dr. Davidoz, who had been associated with the Nurek Dam in Soviet Tadzikistan (A 315 m high Dam in a highly seismic zone of Soviet Tadzikistan), the 260.5 m high Tehri Dam will be quite safe.

So good has been the progress since the THDC took over the project that its Chairman and Managing Director has drawn up a programme of commissioning the first unit of power by 1995. As of now, the project authorities have completed four diversion tunnels, each of 11 m diameter to divert the water of Bhagirathi and Bhilangana rivers, in order to provide a dry river bed for the construction of the Dam. Also the work for four head race tunnels, each of 8.5 m diameter, has been partially completed. The approach adits (tunnels) for the underground power house cavern have also been completed.

EARTHQUAKE OF 20TH OCTOBER 1991 AND TEHRI DAM

In the meantime, when the work for Tehri Dam was under progress, an earthquake of 6.1 intensity on Richter scale hits Garhwal Himalaya badly on 20th October, 1991 and once-again raised a question against its safety and proper site. However, the Dam has been designed to withstand the earthquake of magnitude 8 + on Richter scale. Many scientists and acadamicians have raised a question that afresh survey on safety aspects is needed. While few others are of the view that it is safe and showed a safer behaviour.

According to Prof. C. Prasad (letter to Editor, H.T. Nov. 5, 91) "This earthquake has exploded the myth regarding the seismic stability of the Tehri Dam. A fresh study of the seismicity of the Garhwal region with the latest technology will be worthwhile". He also added that this earthquake appears to have taken place because of the reactivation of Main Central Thrust (MCT) that passes through north Almora to the localities in the vicinity of Joshimath (Chamoli) and north of Uttarkashi. It is 50 km away from the Dam site. Therefore, the claim of THDC that there is no damage to the Dam site may be true. He pointed out that two important major thrust which are active, passing through the Dam area, viz., the Srinagar Thrust (S.T.) passing through the proposed reservoir while the Nayar Thrust (NT) passing through the south side of the Dam. In his opinion "if reactivation takes place along S.T., it will cause landslides on the sides of the reservoir, causing more siltation. On the other hand, any reactivation along the N.T. will damage the Dam itself".

Mr. Balram Jakhar informed the Rajya Sabha on Dec. 6th 1991 that the Govt. has asked the scientist to undertake a review of safety aspects of Tehri Dam Project. According to him earthquake of 20th October, 1991 did not cause any damage to the Dam.

In this regard, THDC quoted in their press release "The safe behaviour of Tehri Dam and other structures which are designed to be safe against a much more powerful earthquake to magnitude 8+. It has once again demonstrated the adequacy of Dam building technology for construction of tail safe high Dam in seismic environment". But Prof. T. Shivaji Rao of Andhra University alongwith a member of the Expert Committee of Union Ministry of Environments and Forest expressed his views against the contention of THDC experts, authorities and Engineers, that the already built up height of over 15 metres above the river-bed out of the 260.5 metre high Dam had suffered no crack and that the Dam is "absolutely safe". According to him this is mis-leading to make an assessment on the basis of quake's impact which is 100 km away from its focus and too on the already built up height which is quite small.

Renowned Geologist of Geological Survey of India (GSI) Mr. P.C. Navani told to newsmen on 25th Oct. 1991 that in the coming 200 years there is no chance of earthquake occurrence. If we consider the observation of Dr. V.K. Gaur which was given during the submission of his report to the working Group, in which he stated that "The Tehri region where no major earthquake has occurred since 1828 (M-7.6) may thus have large residual strains which have accumulated during this period". And "since, earthquake of a magnitude less than 8.0 do not relax sufficient strain, the probability of a major earthquake whose rupture zone may traverse the Dam site is high....", the view of Mr. P.C. Navani looks true. According to Mr. S.P. Singh, Chairman and Managing Director of THDC, Dr. V.K. Gaur told him that if during the construction period Himalayan area experienced a strong tremor and the project withstood it, he would himself certify that there would be no more earthquake in the area for the next 250 to 300 years and hence the completed Dam would be quite safe. In this regard the view of Prof. Rao's is different, he said that the contention of Tehri expert Team's was not valid, as the earthquake, of 6.1 magnitude, can release only a small fraction of the "accumulated stress".

In the view of Dr. Anand Swarup Arya of Roorkee University, the Tehri Dam Project is safe against tremors. The greatest authority of the earthquake, Dr. Jayakrishna, asserted that the Dam would be quite safe in case of stronger tremors. He also said this and other earthquakes were no danger to the Dam at all.

This is a hot debate started just after the earthquake of 20th October, 1991 for ensuring the safety and reliability of the Dam. We are not recommending any definite point as more investigation is needed on the controversial aspects of safety and reliability due to the appearance of this earthquake. We, therefore, hope and request to Government and Experts that they will sit together for reaching to a definite conclusion, so that the decision will be taken on early basis.

THE PROJECT AT A GLANCE

1.	Location	Tehri, Uttar Pradesh
2.	Catchment area	7511 Km^2
3.	Snowbound area in the catchment	2328 Km^2
4.	Average annual rainfall	101.6 cm to 263 cm
5.	Annual run-off on 90 per cent availability.	5.59 MAF
6.	Maximum recorded flood	3800 m^3/S
7.	Adopted maximum flood for diversion during monsoon	8120m^3/S
8.	Type of the dam	Rockfill, clay core
9.	Height of the dam from deepest foundation level	260.5 m
10.	Height of the dam river-bed level	239.5 m
11.	Installed capacity	1000 MW
12.	Firm power	346 MW
13.	Area irrigated	2.7 lakh ha.
14.	Dead-storage capacity	5 MAF-at 720 m above MSL
15.	Outlet for releasing water for irrigation	At 730 m above MSL
16.	Head-Race-Channels for leading water to the turbines.	Four tunnels at 720 m above MSL.

5

Earthquake of 30th September 1993

An earthquake struck Osmanabad and Latur districts of Maharashtra states on 30th September 1993 at 03.56 hrs, its magnitude was measured 6.3 on Richter scale, at 0442 hrs and 0623 hrs mild tremors again struck the region. According to the seismologists the cause of this earthquake could be the constant stress. The death toll was more than 35,000. Worst affected towns are Killari, Killariwadi, Pomadevi, Haregaon, Sarali, Talani, Sarvaral, Sakral and Omarga. Seven other states also experienced the shocks of earthquake. Epicentre of the earthquake was at latitude 18.2 degree north and longitude 76.2 degree east in Maharashtra, Karnataka and Andhra Pradesh border. Epicentre and affected area are shown in Figure. 5.1.

5.1 CASUALTIES AND DAMAGE

Earthquake of 30th September 1993 ravaged both life and property of Latur and Osmanabad districts of Maharashtra State. Due to this more than 35,000 lost their lives. Near about 85% houses were come on the ground along in Killari village. The extent could be understand easily by the perusal of below stated news articles.

5.1.1 OVER 6,200 KILLED IN MAHARASHTRA EARTHQUAKE

Tremors rock 6 other States—Massive rescue operations on

Bombay, Sept. 30 — At least 6,200 persons lost their lives and 10,000 others were injured in a devastating earthquake that rocked parts of Marathwada and western Maharashtra early today morning.

The earthquake also claimed 15 lives in Karnataka while tremors were felt in parts of Kerala, Gujarat, Tamil Nadu, Goa, Andhra Pradesh and Madhya Pradesh.

Massive rescue operations were on with Army jawans and police personnel extricating bodies from the debris in the worst-affected Omarga and Killari townships of Osmanabad and Latur districts respectively.

The earthquake, which measured six on the Richter scale, took a tool of nearly 3,000 lives each in Omarga and Killari towns, according to Maharashtra Chief Secretary M. Raghunathan.

Thousands of houses in Latur and Osmanabad districts were reduced to rubble. Houses and buildings crumbled in less than 20 seconds when the main tremor was felt.

Seven more tremors followed but those were of lesser intensity. People believe that there may be another quake after 24 hours and this has caused deep panic among them.

Maharashtra Chief Minister Sharad Pawar, who rushed to Latur, told newsmen there that 25 companies of the State Reserve Police, 50 companies of the Army have been requisitioned to tackle the worst disaster in the history of Maharashtra.

A large contingent of medical team is also arriving to provide relief to the victims.

The villages worst affected are Talani, Killari, Killariwadi, Pomadevi, Haregaon, Sarali, Sarvaral and Sakral.

The condition was so pathetic that people were starving due to want of food.

The quake also wreaked havoc in Nasik, Osmanabad, Pandharpur, Vai (Satara), Solapur, Ahmednagar and the adjoining villages of Killari and Omarga, claiming another 200 lives.

Eighty per cent of the houses in Killari and 21 neighbouring villages having a population of 40,000 collapsed and 60 per cent of the houses in Omarga and adjoining 20 villages with a population of 40,000 were destroyed in the calamity, official sources said.

Thirteen columns of the Army were rushed from Bombay along with engineers from the Army's Bombay Engineering Group to assist the State administration in relief and rescue measures.

The Army's sub-group from Bombay has rushed 10 medical units to the quake-hit areas. Besides mobile hospitals with 42 beds and 10 medical teams from the Bombay Municipal corporation have also left to help the victims. In addition, 20 civilian medical units from Delhi and six others from Pune with 600 bottles of blood were air-lifted to the affected areas.

Maharashtra Finance Minister Ramrao Adik said a sum of Rs. 10 crore had been allocated for relief operations and added that Rs. 25,000 would be paid to the next of kin of the deceased. The Mayor of Bombay Mr. R.R. Singh, has announced disbursement of Rs. 1 crore from the Mayor's fund.

Officials said access to some villages in Latur and Omarga had been cut off as the recent heavy rains had damaged roads, making relief efforts difficult.

The toll figure was high apparently because most of the houses were built of mud and stones, according to State Director-General of Police Shivjirao Baraokar, who left Bombay for the spot in the evening after co-ordinating various measures.

He said police had sent an SOS to unit commanders of the police of neighbouring States to rush their officials and jawans to the scene of the tragedy with necessary equipment to help the victims.

There was, however, no damage to any of the numerous dams in the area and highways and roads also remained unobstructed. There have been no reports of damage to the two famous temples dedicated to Lord Shiva either, Chief Secretary M. Ranganathan said.

Mr. Ranganathan appreciated the tremendous support he had received from the Chief Secretaries and other senior officials from neighbouring States, including Madhya Pradesh and Andhra Pradesh.

While Mr. Ranganathan had appealed to the MP Chief Secretary for a contribution to the Maharashtra Chief Minister's Relief fund, the AP

Chief Secretary had been requested to rush aid to the two districts from Hyderabad which was geographically closer than Bombay and could be reached by road in four hours.

Meanwhile, the Revenue Secretary of Andhra Pradesh has offered to donate 12 ham network wireless communication sets to Maharashtra, 10 each of which will be set up in Killari and Omarga and two in the Maharashtra Secretariat building for direct communication, Mr. Ranganathan said.

The American consulate in Bombay and the local unit of the United Nations International Children's Emergency Fund (UNICEF) had also offered help, he said.

Union Home Minister S.B. Chavan has rushed to Latur adjoining his hometown, Nanded, to supervise the relief operations. Lok Sabha Speaker Shivraj Patil, has also headed for Latur, his constituency.

BANGALORE: Till late this evening, the toll in Karnataka in the earthquake was 15 with more than 60 injured. While seven people died in a house collapse in Indi sub-division of Bijapur, a two-year-old child was buried alive in Basvakalyan taluk of Bidar. Two other deaths were reported from Raichur and six in Gulbarga district.

Official sources in Bangalore said many houses collapsed as a result of the high intensity earthquake, which, according to the Seismic Laboratory at Gouribidanur, measured 4.9 on the Richter scale.

While a portion of the famous Gulbarga fort also suffered some damage with the main entry gate caving in, cracks also developed in a side wall of the world renowned Gol Gumbaj complex. Eight house collapses were reported from Bidar. Some damage to property was also reported from Raichur.

In Bangalore, people were shaken out of their slumber at around 3.55 a.m. The tremors which continued for several seconds, were accompanied by a rattling sound that forced people out of their houses.

Gouribidanur officials said this was the biggest earthquake in the country in recent memory.

The earthquake toll in the State was highest since the Koyna earthquake of De. 11, 1967 when over 100 persons were killed.

Chief Minister M. Veerappa Moily, who in Delhi for talks with the Congress high command, flew to Bidar in a special Indian Air force plane along with Union Home Minister S.B. Chavan.

THIRUVANATHAPURAM: Mild tremors were felt in many places in northern Kerala today morning. According to the reports reaching the State capital, the effect of the quake was felt in Kasaragod, Kozhikode, Malappuram and Thalessery.

However there were no reports of any damage to property or loss of human life till this afternoon.

According to the Meteorological department spokesman here, the epicentre of the quake was recorded approximately at 1000 km north of Trivandrum.

The intensity of the tremor in Kerala was mild. It would have gone unnoticed but for the rattling of window panes.

GANDHINAGAR: The shocks of the massive earthquake which hit Maharashtra were also felt in Surat and Broach districts of south Gujarat.

According to reports reaching here, panic-stricken people rushed out of their homes when they felt their doors and windows rattling and houses trembling.

There were no reports so far of any loss of life.

The seismologists had not anticipated the devastating earthquake in which several thousand persons died this morning near Latur district of Maharashtra. However, stress has been constantly accumulating for a long time around that area, according to the spokesman of the seismology section in the meteorological office here.

The earthquake is of moderate intensity having its magnitude 6.0 on Richter scale with epicentre at latitude 18.2 degree north and longitude 76.7 degree east in Maharashtra, Karnataka and Andhra Pradesh border.

The epicentre is located about 100 kms north-east of Solapur, about 400 km from Bombay and about 10 km from Latur. The cause of the earthquake could be the constant stress. According to officials of the seismology section the earthquake of moderate intensity occurred due to release of stress which has been accumulating for a long time.

There has been slippage of rocks between the Godavari faults and Koyna region. There have been four earthquakes in the peninsula part between latitude 14 degree north and 22 degree north, and longitude 73 degree north and 81 degree east.

Earthquakes of magnitude 6.0 and above on Richter Scale have been occurring since the latter half of 1700. The spokesman of the seismology section they recorded an earthquake of manitude 6-0 on Richter Scale in Mahabaleshwar in 1764, one of magnitude 6.0 in Bilary in 1843, of magnitude 6.3 in Sathpura in 1938 and in Koyna in 1967 of magnitude 6.5 on Richter Scale. In the Koyna earthquake 177 people had died.

The seismologists are of the opinion that earthquakes occur on the basis of focal depth, slippage of rocks or in other words mechanism of movements of rocks. They are devastating in those regions where the number of population in quite large especially if they occur at nights early mornings when there is no activity in the region.

Regarding this particular earthquake, the spokesman of the seismology section said it has occurred due to constant stress by continuous movement of rocks towards north and north-east and collusion with Eurasia, affecting the border areas. They said that wherever there is any weak point earthquake occurs.

In fact, seismologists stated, on 6 Aug. 1988 an earthquake of magnitude 6.8 occurred on the Manipur-Burma border in which three people died whereas on Aug. 28 the same year, a quake of magnitude occurred on Bihar-Nepal border killing 1,004 persons.

In 1992, Uttarkashi earthquake on Oct. 20 was of magnitude 6.6 and earthquake at Jaisalmer on Nov. 8 was of magnitude 6.4 on Richter scale.

Earlier also earthquakes occurred as it did on July 7, 1980 when as many as 200 persons died in Dharchula. In Jammu 15 persons died on Aug. 8, 1980, in Cachar earthquake 11 persons died in Dec. 1984 and in Bangladesh earthquake which was of a magnitude 6.1 in 1989 one person died, said the seismologists.

In Latur district an earthquake had occurred last year also. It was on Oct. 18 when the district was rocked. The magnitude of the earthquake was 4-5 on Richter Scale.

Courtesy : The Hindustan Times

5.1.2 QUAKE TOLL CROSSES 21,000

Rescue work on war footing

Killari (Maharashtra), Oct. 1 — Army and police mounted massive rescue operations in the quake-hit areas removing debris with cranes to rescue hundreds of victims trapped in the flattened buildings.

Funeral pyres burnt all over the fields of Killari and Omarga as relatives cremated over 21,000 victims of India's worst earthquake disaster in five decades and Army and civil rescue teams rummaged through the rubble to look for survivors.

The number of perished may cross 28,000, according to information from various quake-hit areas and relief and rescue teams.

Mass cremations of victims took place amid the wails of the wounded survivors and the kith and kin of the deceased in the villages, where the earth swallowed up thousands of people asleep in their homes.

Officials estimated the bodies recovered at around 12,000 with about 11,000 others escaped with injuries in the midst of the continuing rescue efforts to dig more people out of the debris.

Medical teams have arrived to provide relief to the victim. Medical aid and assistance were pouring in from all over the country and abroad, official sources said.

Rescue operations were disrupted last night after power and telecommunication links snapped and a heavy downpour lashed the affected villages.

The hurried disposal of the dead carted away for burial and cremation was promoted by fears about an outbreak of epidemics from the decomposing bodies lying in heaps in the rubble, authorities said.

The stench from the decaying bodies entombed beneath the flattened houses was overpowering, reflecting the magnitude of the toll and raising apprehensions of further health hazards.

Operation Sahayata, the Army's relief and rescue efforts, has been stepped up and upto 80 Army columns had been already deployed.

The Army's work had been projected well into the interior. The mechanised infantry regimental centre at Ahmednagar had established a relief management headquarter to coordinate relief work in remote areas.

Specialised engineer task forces have also moved out with earth moving and recovery equipment. The Army had stepped up medical aid, deploying three field ambulance units capable of functioning as temporary hospitals.

Six surgical teams had been deployed. The Additional Director of Medical Services at Southern Command Headquarters, Pune had also moved to the earthquake-hit districts with surgical teams, with 230 packages of medicines.

The Army has provided substantial tents at the disposal of the civil administration. Cooking ranges have also been provided. Close liaison is being maintained with the civil administration by Army authorities to provide immediate response.

Fifteen thousands troops have contributed their one day's rations, which were being rushed to the affected areas.

Maharashtra Chief Minister Sharad Pawar who is camping at Sholapur since yesterday morning, has directed officials to immediately set up sheds for temporary shelter to the people rendered homeless. The facility, which includes medical aid, food supplies, electricity and water, would continue for at least 25 days till alternative and permanent arrangements are made.

Efforts are already underway for permanent rehabilitation of the quake-hit people. The Reserve Bank of India (RBI) has directed all commercial banks to expedite all necessary financial assistance to the affected people.

Mr. Pawar said Government would consider shifting and rehabilitating the survivors of the earthquake in Latur and Osmanabad districts at safer places.

The Chief Minister said adequate relief had poured into Sholapur the staging post for the massive relief work being undertaken by the State Government with the assistance of military and para-military personnel. He said food, clothing medicines and temporary shelters were being provided on a war-footing to the 129,000 people rendered homeless.

Mr. Pawar is regularly visiting Killari and Omerga.

State Chief Secretary N. Raghunathan, who was in constant communication with Lt. Gen. A.S. Kalkat, General Officer Commanding in Chief, Southern Command, told reporters at Bombay that the first priority was to extricate survivors from the debris.

Mr. Raghunathan said he had been informed by Lt. Gen. Kalkat that 40 bodies were recovered from Killari this afternoon.

He said the rescue work was hampered by some rain today but the Army had moved in its bulldozers and more had been requisitioned from the State Irrigation Department.

Asked whether the toll would rise further, the Chief Secretary said the estimates put out by newspersons and independent sources were at best guesses". However, he admitted that in 29 villages of Omerga taluka alone, the Army had managed to clear the debris only in 13 villages and work of removal of debris was continuing in 16 other badly affected villages.

Maharashtra Chief Secretary N. Raghunathan said that 30 villages in Omerga region of Osmanabad district and 25 villages in Killari region of Latur holding a total population of 1.25 lakh had been affected by the earthquake. Another 27 villages suffered minor damages.

Several industrial houses placed their private aircraft at the disposal of the Government for relief work. Similarly, private air-taxis such as the East-West Airlines are lifting relief supplies to the quake-hit areas.

Five star hotels are contributing thousands of "paratha packets" for relief work. Textile mills have contributed dhotis and clothes and Hindustan Lever has directed its dealers in Sholapur and areas around to place all their stocks of tea, soaps and edible oils at the disposal of the authorities for distribution among the people in distress.

Both white and blue collar workers the State Government and private sector have donated liberally to the Chief Minister's Relief Fund. Political parties suspended their agitations and assured full cooperation to the Government. The BJP has suspended collection of funds for the party.

The Reserve Bank of India asked all commercial banks to expeditiously extend relief and rehabilitation assistance to the affected persons.

Killari village sarpanch S.D. Parsagir said there were little chances of any survivors in villages like Talni, Killari, Wada and Manglur. The death tool may go up to 35,000, he said.

A survivor from Talni village said 32 members of his family were killed.

Latur and Osmanabad fall in the backward region of Marathwada in south-east Maharashtra. While Latur has a narrow gauge railhead, Osmanabad is not connected by rail, though the railway track to Latur passes through the district. People here are mostly poor farmers living in bamboo and mud houses which easily collapsed during the tremors, accounting for the high number of casualties in the tragedy.

One battalion of the light infantry was assisting the State Administration working directly under six special secretaries at the site of the tragedy. More than 20 companies of the State Reserve Police were assisting the authorities and Director General of Police S.V. Baraokar was also on the spot.

Bharatiya Janata Party (BJP) leader L.K. Advani, who toured some of the worst-affected villages along with party general secretary Gopinath Munde, told newsman at Latur that there was evidence that the Government machinery had failed to do its job. The need of the hour was to take up work on a war-footing, which was not being done, he alleged, and said that the bodies lying unattended had started to decay.

The Western Railway had also rushed supplies to Latur and Osmanabad and was considering sending special trains to the area if the demand arose.

Courtesy : The Hindustan Times

5.1.3 RESCUE OPERATION STEPPED UP AS TOLL TOUCHES 30,000

From Ashok Das and Agencies

Latur (Maharashtra), Oct. 2—Bodies piled up along roadsides today as rescuers used cranes, bulldozers and even their bare hands to search through mounds of rubble for survivors, two days after a killer earthquake left more than 30,000 people dead in a death dance in Latur and Osmanabad districts.

Four victims of the quake were found today in a state of shock. One in Satur and three in Tuashigarh. They included two children aged six and four. Another man was extricated from the debris in a dazed condition at Uditpur.

Fig. 5.1.1 Family members wail as bodies of quake victims are salvaged from the rubble at Killari village

Courtesy: Hindustan Times

Indian Army Sappers found limbs of the dead sticking up from the debris in Killari village that bore the brunt of the worst national disaster in 50 years even as overnight rains disrupted their largest peacetime operations.

Indian Air Force (IAF) today pressed into service nine Chetak and MI-8 helicopters to supply food to villagers here and the surrounding earthquake-hit areas.

Even as mass cremations continued for the third consecutive day today, IAF choppers were flying in here from Bidar air base in Karnataka at a makeshift helped here.

A new wave of panic spread through the survivors camping in open fields as mild tremors again shook Latur and the surrounding areas for the second successive night but no fresh damage has been reported so far.

More than 2,000 doctors from across the country are tending the injured and police has forced a rapid action force to bring relief to the affected areas where 21,500 people have died so far according to the official count.

Fifteen medical teams from the naval hospital "INHS Ashvini" Bombay, are tending the wounded. Three field ambulances manned by 40 doctors and 100 paramedics and equipped with surgical and blood transfusion gadgets have fanned out to 14 villages, including Killari.

Thousands of bodies in various stages of decomposition recovered from there were either cremated or buried en masse to prevent outbreak of epidemic, health officials said.

Meanwhile, a report from Sholapur said search for survivors was called off in many affected villages of Osmanabad and Latur districts from today evening even as local people and relatives complained that thousands of people remained trapped under the debris.

Col. Roy of the Army, supervisioning the overall operations in Osmanabad district, told The Hindustan Times that it had been decided to end search operations in seven affected villages of Osmanabad with effect from today evening. He said the decision was taken in consultation with the local sarpanches.

The Army's decision has thrown a pall of gloom on the relatives of the victims, who were hoping against hope that some of the victims might still be alive.

Col. Roy said from tomorrow the Army would concentrate on relief and rehabilitation work for the survivors.

"Operation Search" for the survivors had not been very fruitful. During the past 36 hours of joint operations by the Army and Maharashtra police, nearly 700 bodies were recovered from the villages of Sastur, Holi, Pet Sanghvi, Ekaundi, Rajegaon, Udatpur and Kaithan.

There were, however only four survivors which comprised three children and one old man, the Army said. In neighbouring Latur district, search teams rescued only three people from the debris.

The stench of dead bodies and carcasses of buffaloes and cows was overpowering. In Peth Sanghvi village, teams of Army Personnel were busy digging out bodies and lighting funeral pyres. Survivors were seen collecting whatever they could retrieve from their devastated homes. A group of three women were seen standing near a mound under which their seven family members were buried.

The Army personnel told this reporter that today morning they dug out a two-year-old girl from the debris of one of the houses. The child was in great pain as both her legs appeared to be have been crushed under the weight of the boulders. The 'jhula' in which she was sleeping at the time tremor struck apparently upturned on her and protected her from being ourshed.

Maharashtra Chief Minister Sharad Pawar admitted that the Government was facing a "daunting task" even as donations, medicines, foodgrains and offers of aid poured in from all over the country and abroad.

The Maharashtra Government has appealed to donors to desist from contributing in kind and instead make cash donations to the Chief Minister's Relief Fund. Finance Minsiter Ramrao Adik said in Bombay that the Government was "overwhelmed" by the large amount of the foodgrain, sugar and building material that had been sent for the relief efforts in the affected areas.

The heavy downpour quite unusual at this time in the districts has caused waterlogging, hastening the decay of bodies. Disposal of the bodies has become a problem with dry firewood being in short supply. Hundreds of bodies were piled up four cremation along road sides in Killari, Sastur and other villages.

Mr. Pawar said the government's first priority at the moment was to provide succour to the affected people. The need of the hour was to provide, on a priority basis, food, medicines, drinking water and temporary sheds for shelter.

"We have to protect the lives of those who have survived and whatever is possible will be done on a war footing", the Chief Minister said.

Mr. Pawar said his endeavour would be to build new houses with facilities for markets, schools and other civic amenities in every affected village. Wireless communications had already been established in every village by the Army and the collectorated of Latur and Osmanabad.

He said a master plan would be prepared for construction of quake-proof houses with the advance of experts.

Courtesy : The Hindustan Times

5.1.4. ONLY STENCH OF DEATH REMAINS

Killari Oct. 2 (PTI) — A solitary water tank stands high bearing testimony to the existence of a village called Killari. As dusk set in on Wednesday the 15,000-odd inhabitants of the village went to sleep after a hectic night of Ganesh immersion, little realising that the dawn would only bring devastation and death.

Come Thursday and the village to reduced to a shambles with the all-pervading stench of death spreading all over. The devastating earthquake that hit the area in the wee hours of Thursday had claimed the lives of more than half the population, most of them still buried under the debris.

It's a pathetic sight, comments an Army personnel, pointing out to a pile of mangled bodies — men, women and children - waiting to be cremated. The bodies are all wrapped in whatever piece of cloth the relief personnel could lay their hands on.

By Friday noon 2,000 bodies had been extricated, cremated or buried en masse. In fact, the first sight one encounters on taking the minor road leading to the village is one of burning pyres and a huge pit dug up for mass burial.

Area tahsildar D.G. Phulari, having gotten over the initial numbness following the heavy loss of life, is now busy chipping in the coordination of relief work, which really picked up only yesterday with the arrival of men and equipment from the Army. At least 8,000 bodies are still buried under the debris in Killari alone and there are eight other smaller villages where the situation is similar, he said.

He goes on to explain that most of the inhabitants were of the labour class, strangely though only 15 per cent of the surrounding area is under cultivation.

As one moves deeper into the village, the scene is that of quarry with granite blocks barring every way one goes.

Perched atop one of the granite piles is desolate 19-year-old student of Latur medical college. Srinivasa Nokja is yet to recover from the shock of losing his teacher-mother and brother and his father and sister hospitalised with serious injuries.

Srinivas himself escaped death since he stays in a hospital at Latur, points out to what was once a "Kirana" shop owned by his family. A small upturned showcase containing some consumables are all that indicate that this place was a shop "When I landed here, I found the almirah broken and some gold ornaments missing", says Srinivas. Was it looted or is buried under the debris? Only time will tell.

Further up a steep gradient, Army jawans and personnel from the Engineering Regiment from Secunderabad are busy clearing up a large area of debris. Covering about 500 sq. m. all the houses in this area have been flattened, leaving no chance of survival for the inhabitants of the 70-odd houses here. Entire families wiped out, and scores of cattleheads too.

From the top of the village on the gradient, one sees a sea of humanity pouring into Killari from all sides. Voluntareers, politicians, police, Army, and Media people merge against the backdrop of the milling crowds, most of whom want to catch a glimpse of the effect of the biggest ever earthquake to hit this side of the country.

Says Tara Devi, a social worker from neighbouring Latur town, this is not a mela for people to flock in without any objective. "I and three other social workers have been persuading the pouring masses to participate in the relief operations since the magnitude of the calamity is such that every hand would matter."

The magnitude of the calamitous earthquake on the fateful night of Thursday was so intense that cracks appeared on the parchased earth as wide as 5 to 6 feet and 12 and 13 feet long in a radius of 100 kms area.

All that was left after the devastation were crumbled houses with bricks, stones, television sets, bicycles and divine pictures strewn all over.

Almost all the animal stock of 52-odd villages and several families in the quake-hit area were swallowed alive by the earth which trembled for some time, one of the survivors said.

The scene was pathetic with only limbs and faces of some persons visible on the surface of the ruins. Thick layers of soil from underneath the earth spewed out in the impact of the calamity.

While scores of people, including women and children, were still buried under the debris, bodies of victims were piled up on the roads and fields for mass cremation. A bird's eye view from atop the ruined houses gave an impression that the dead were still alive lying in deep slumber in their dwellings.

The quake did not even spare those who were houses as villagers were killed by heavy boulders falling on them.

Most of the victims were Hindus and Muslims with an average population of about 2,000 in each village of the total 52 affected, eyewitness accounts said, adding that the bodies have started decomposing. The rescue work was hampered following heavy rains last night.

Panic gripped the people in quake-hit areas as Latur experienced mild tremors last night. Some of the survivors have fled the scene while people from surrounding villages flocked to the affected areas to help the State administration in rescue operations.

The barking of dogs as also crowing of cocks — a normal feature in these sleepy villages was now not heard. Stables of cows and buffaloes wore a deserted look as the entire livestock of these villages have perished.

Courtesy : The Hindustan Times

Fig. 5.1.2 Bodies in front of the makeshift medical centre at quake-hit Killari

Fig. 5.1.3 A grief-stricken woman, the lone survivor of her family, at Chicholi village in Latur district.

5.1.5 NIGHTMARE OF QUAKE RECEDING

Ashok Das

Omerga (Osmanabad) Oct. 4 — Hundreds of tragedies are unfolding in the earthquake hit areas of Latur and Osmanabad districts. The nightmare of the black Thursday having receded into the background, the survivors are slowly realising the enormity of the situation. While entire families have been wiped out, the lucky survivors have lost someone or other. They now feel that the dead are the blessed for they do not have to live with mental and physical scars and worry about an unknown and uncertain future.

The quake has particularly been unkind to the scores of old men and women and children. While the old people, dependent on their able bodies children for sustenance have nowhere to go whereas the children particularly the elder ones, are terrified of what the future holds for them.

Lying in bed number 14 at the Vivekananda hospital at Latur is Geeta Bai. Even as the doctors are busy treating her for a possible fractured arm and a head injury she sobs intermittently. She is least bothered about what happens to her. She sits brooding over her husband, sons, daughters and grand-daughter, 13 in all, who have perished in the earthquake. She refused to touch the fruits and bread left by someone near her.

A few feet away, Mahadev Rao of Mangrur village, lies on the floor with multiple injuries and concussions on the head. He is the sole survivor of what was before Sept. 29 a happy lower middle class family. His entire family of six members were killed in wall collapse. He survived as he was sleeping outdoors.

Ravi Shankar of Limbala village has escaped with a gash on the neck and few bruises elsewhere. He has lost his wife, two sons and a daughter. He says he wants to die as there is nothing a left on the world.

The child ward of the hospital is overflowing. Unless the patients are serious enough to need hospitalisation, the doctors are treating them as outpatients. One of the notable patients in the ward is a 15-day-old girl, now christened as Nivedita by the nursing staff of the Vivekananda hospital. She was brought to the hospital by some local people who found her crying amidst debris close to her mother's deadbody. Her entire family is presumed dead and no relative has bothered to claim her. One of the nurses at the hospital has decided to adopt her. She is now fed milk and liquids through intravenous pipes. Doctors say she had not cried even once since was brought from the debris.

There is another two-year-old girl being treated at the Government hospital at Omerga. Doctors are treating her for fracture of both legs and she has not opened her eyes, partly because of pain and partly because of sedation. These two young girls are too young and may not know the truama of separation from their natural parents but not all are lucky as them.

Ravi, a young boy of eight years is being treated in the some hospital. He has none left in the world. He is too dump found to speak. Labnya, a girl of 10 years has lost her entire family. She escaped without any injuries but her family of parents and two sisters died in the tragedy. She may have to stay with her uncle. She does not know what future holds for her.

All the hospitals are being mobbed by the family members on relatives of the missing with the fond hope of finding their relations alive. Most have to return disappointed. But still quite a number of them squat in and around the hospital with fond hope that their missing relative may be brought in by one of the hundred odd ambulances.

Courtesy : The Hindustan Times

5.2 RELIEF AND RESCUE OPERATIONS

Relief and rescue operation started just after the first news of destruction. Initially the pace was slow due to the high extent of distribution, and damaged communication system. Bad weather also hampered the pace of rescue operations. Indian military deployed for this work did hard work to overcome this problem. Mass cremation of dead bodies was done. News articles appeared in Hindustan Times are self explanatory in this regard.

Courtesy : The Hindustan Times

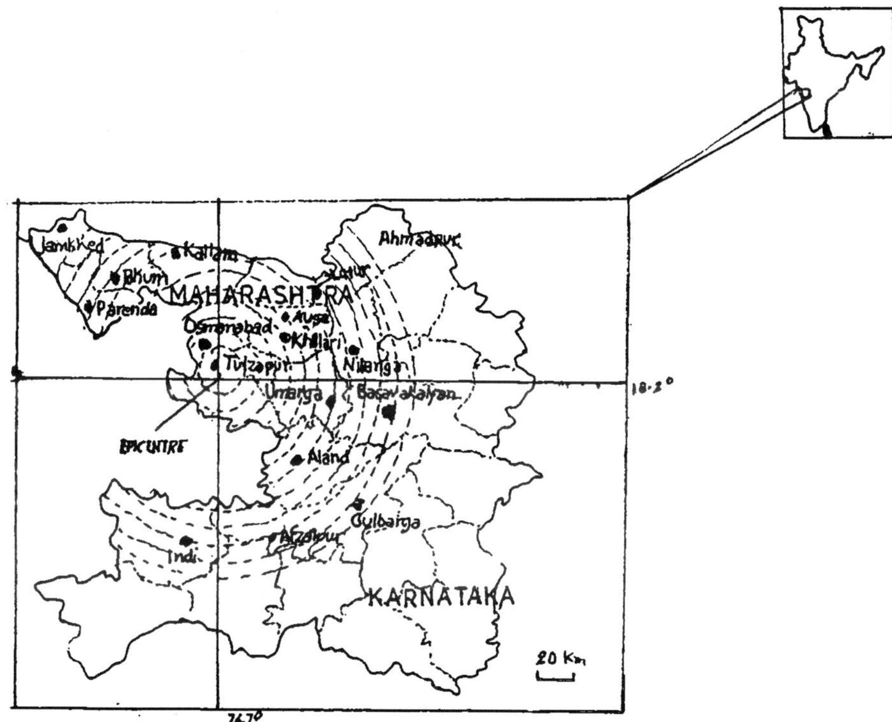

Fig. 5.2 Worst Affected Area

5.2.1 RELIEF BEING SENT TO AFFECTED AREAS

New Delhi, Sept. 30 — A Crisis Management Group headed by the Cabinet Secretary Zaffar Saifullah was convinced today to mobilise supportive activities of the Central Government agencies and the Army to provide assistance to the earthquake-hit populace.

Five medical units have been rushed to the spot, while a massive mobilisation of doctors, paramedics and medical supplies are awaiting transport tonight.

The Relief Commissioner and Cabinet Secretary were in constant touch with the relief headquarters in Bombay to assess the developing situation and are organising personnel and relief supplies accordingly.

Several earth moving equipment and recovery vehicles are being rushed to the quake-hit area to clear the debris and look for survivors.

Voluntary organisations have organised several community kitchens to provide food for the homeless.

The Sholapur and Osmanabad region has suffered heavy damage. Latur and Killari were the worst hit, with the earthquake intensity ranging between 5.5 and 6 on the Richter scale. The tremors began at 3.30 a.m. and went on till 6.45 a.m.

In place, the ground has experienced an upheaval and whole towns have collapsed.

Home Minister S.B. Chavan and Lok Sabha Speaker Shivraj Patil, visited the areas around Satara, while Maharashtra Chief Minister Sharad Pawar rushed to Khillari.

According to report reaching here, the entire revenue administration of Maharashtra has been mobilised. Damage of varying intensity have been reported from Bidar, Bijapur and Raichure.

Agriculture Secretary Dr. M.S. Gill today sought immediate release of Rs. 11.5 crore to the Calamity Relief Fund for the Maharashtra Government.

This will be in addition to the two crore rupees released to the Maharashtra Government today from the Prime Minister's Relief Fund.

Meanwhile, an official spokesman said, on a directive from the Defence Ministry, Army personnel had been moved from the Secunderabad, Pune and Ahmednagar to the quake-hit areas to assist in rescue and relief operations.

Tents have been provided to give shelter to those rendered homeless and injured. The Union Health Ministry is also sending its teams to the affected areas, he said.

The Andhra Pradesh Chief Minister K Vijaya Bhaskara Reddy today sanctioned Rs. 10 lakh.

Courtesy : The Hindustan Times

5.2.2 MAKESHIFT CREMATION SITES

Latur, Oct. 1 (PTI, UNI) — Mass cremations and burials were being carried out near the devastated sites of quake-hit villages and in the open field for the second consecutive day today.

Fig. 5.2.1 Several people watching the mass funeral of quake victims in Killari village on Sunday

Relief workers were carrying the bodies to makeshift cremation sites to dispose of the bodies which were extricated from the debris.

A team of PTI correspondents, which toured the affected villages, witnesses bloated bodies of men, women, and children all over the places, especially Killari and Tarini in Latur district and Kawatha, Rajagaon and Sastur in Omarga tehsil of Osmanabad district.

The exact number of dead would be known only after the rubble of thousands of mud-built and concrete structures is cleared, police sources at Killari, 45 kms from here said.

Several companies of the Army which arrived early today and SRP jawans, police personnel including those drawn from Bombay, were assisting in relief work. However, a heavy downpour that lashed the affected villages hampers the rescue operations.

Three new-born babies were today rescued from the debris of the Killer earthquake that hit the district yesterday.

A 15-day-old infant was rescued from the worst affected Killari village and was admitted to the Vivekananda Hospitals here. Doctors attending on the child said the parents of the new-born child were missing and believed killed in the quake as some voluntary workers had brought the child to the hospital.

A Rashtriya Swayamsevak Sangh (RSS) worker also rescued six-year-old boy trapped under the debris for 30 hours at Satur, another worst hit village.

A three-month-old baby girl, who has not even been christened, and whose parents perished in the quake has been placed under the care of doctors at the Public Health Centre at Killari.

According to villagers with whom the correspondents spoke to in Killari, Rajgaon, Kawatha, Tarini, Ekundi and Nagangwadi, between 60 to 70 per cent of the population of each village had perished in the calamity.

Many of those who died or are trapped under the stones and mud, are women and children. In Killari, one could see a number of stray dogs and pigs rummaging through the debris. In Tarini village having a population of 5,000 people, only 500 survived, villagers said. Similarly in Rajgaon 600 out of population of 1,000 died and every single house was in ruins.

Most of the residents in the 30-40 kms radius of this quake-hit area are farm labourers.

Courtesy : The Hindustan Times

5.2.3 WORLD LEADERS SEND CONDOLENCES

Undated, Oct 1, (UNI)— United Nations Secretary General Boutros Boutros - Ghali and a host of world leaders have expressed their concern and distress over the heavy loss of life and property in yesterday's earthquake and pledged financial assistance in an international effort of solidarity with the quake victims.

A spokesman for the UN chief said Mr. Boutors-Ghali had expressed his sincere condolences to the Government and people of India for the "appalling loss of life and devastation caused by the earthquake.

"The United Nations is already responding to this disaster and stands ready to assist the Government in its relief efforts. He is in personal touch with Prime Minister Narasimha Rao", the spokesman added.

UN General Assembly President Samuel R. Insanally, in his message, extended his "deepest sympathy" to the quake victims and expressed the hope that the international community would respondent generously to any requests for assistance.

Mr. Sergei Shoigu, Chairman of the Russian State Committee for Emergency Situations, said a team of rescuers and doctors of the Russian centre 'Catastrophic Medicine' was on standby at a Moscow airport to fly to the quake hit areas.

The highly professional and most modern equipped rescue team was awaiting a formal Indian request for help.

Queen Elizabeth II England, in a message to President S.D. Sharma, expressed her profound concern over the calamity and said, "the Duke of Edinburgh joins me in sending our sincere condolence to you and to all those families who have suffered.

Prime Minister John Major in a message to the Indian Prime Minister said, "I wanted to you know that our heartfelt sympathies are with all those whose lives or families have been devastated by the earthquake. We stand ready to help in whatever way we can."

A report from Tokyo said the Japanese Government would send $ 25,000 in aid while the Japanese Red Cross Society announced giving an emergency aid of $ 52,380 aid providing medical aid to the victims.

German Chancellor Helmut Kohl, in message to Prime Minister Mr. P.V. Narshimha Rao, expressed his Government's sympathy heartfelt for the victims and announced giving an assistance of Rs. 20 million for rehabilitation and reconstruction purposes.

German Foreign Minister Klaus Kinkel in his message to External Affairs Minister Dinesh Singh, said, "it was a great shock for me to hear about the natural disaster which has struck your country."

French Prime Minister Edouard Balladur, in a message to Mr. Rao expressed his "deepest symapthy" and said the French government would be ready to participate in the international effort of solidarity.

The French Government would contribute Rs. 30 lakh to the Prime Minister's Relief Fund for the rescue operations in Maharashtra. It said the French specialised technical teams were ready to fly to India to take part in emergency rescue operations.

Australian Prime Minister Paul Keating, in a message to Mr. Rao, conveyed the "dee sympathy" of the Australian people and said, "The quickest and most efficient was for Australia to assist in disaster relief in this instance is through a cash grant."

Mr. Keating said Australia would provide 2,50,000 Australian Dollars through the United Nations Department of Humanitarian Affairs (DHA) as a first response.

Bangladesh Prime Minister Beguam Khaleda Zia in her message, extended "heartfelt condolences and sympathy" to the members of the bereaved families.

"It is with deep sense of shock and grief that we have learnt about the disastrous earthquake resulting in death and devastation on a massive scale, the message said.

President Kim Young Sam of South Korea, in his condolence message to Mr. Rao, expressed his shock and distress over the calamity and said, "I earnestly pray for prompt rehabilitation and speedy return to normalcy."

The Council for the Heads of Arab Missions in New Delhi in a message said, "We are shocked by the devastating earthquake and the resultant human suffering and tragic loss of life and property in the Marathwada region and other areas."

The message added, "We wish to contribute for the alleviation of their human sufferings."

The Italian community in New Delhi, through the Embassy of Italy, expressed its solidarity to the families "Aggrieved by the earthquake" and pledged an initial collective contribution of Rs. five lakh to the Prime Minister's Relief Fund.

President of Sri Lanka D.B. Wijetunga in a message to President Dr. Shanker Dayal Sharma and Prime Minister P.V. Narasimha Rao, said he was "deeply distressed by the tragic loss of life and destruction caused by the devastating earthquake," yesterday.

The British opposition Labour Party, expressing shock at the heavy death toll in the quake has urged the British Government to render assistant to tackle the situation.

Labour leader John Smith sent a message of sympathy to the Indian High Commissioner L.M. Singhvi, while labour spokesman, Jack Cunningham opened the foreign policy debate at the party conference in Brighton with reference to the huge loss of life in the earthquake.

The British High Commissioner in New Delhi today donated £ 25,000 towards the Prime Minister's relief fund.

The Indian community in Britain has launched a relief fund campaign to help the quake victims.

Several voluntary organisations, besides the ethnic radio and TV networks have issued appeals for generous donations to the relief fund. The British Red Cross and other local charities too, are lending support for the humanitarian cause.

Special congregations were organised at places of worship in different parts of the country last night where people of Indian origin prayed for the earthquake victims.

The Indian High Commission here has opened a helpline for the Indian community in the UK to know about the well-being of their relatives living in the areas hit by the earthquake.

Australia will provide 3,20,000 US dollars for emergency earthquake relief in India. Development cooperation Minister Gordon Bilney said on Friday.

Iranian Foreign Minister Ali Akbar Velayati has expressed deep sorrow over the loss of the life caused by the earthquake in Maharashtra.

In a letter of condolence to his Indian counterpart Dinesh Singh yesterday, Mr. Velapati said, "I deeply regret the huge loss of life in India caused by the quake. While expressing my deep sorrow over the incident, I feel great sympathy for you, families of the fallen victim and the survivors of the disaster, Iranian news agency IRNA reported.

Courtesy : The Hindustan Times

5.2.3 CLINTON RUSHES AID FOR VICTIMS

Bombay, Oct. 2 — US President Bill Clinton has directed his administration to immediately rush two air-craft loads (200 truckloads) of relief materials for the earthquake hit areas of Maharashtra.

The first aircraft is expected here tomorrow and the second a day later. The relief materials include medicines, food, tents, jerycans etc.

Convoys of trucks are being lined up at the Sahar International airport under the charge of a Minister, to pick up these supplies as soon as these land and rush to the needy areas."

A World Bank team is expected to be here in the next few days to assess the damage caused by the earthquake and the resettlement equipments.

Disclosing this here today evening Maharashtra Chief Secretary N. Raghunathan claimed that 72 per cent of the rubble-rummaging operations have been completed and the work is expected to be completed by tomorrow evening.

Mr. Raghunathan said that baring shortage of 50,000 corrugated sheets, adequate relief materials, both in terms of men and materials have reached the affected areas. Medical facilities, medicines, doctors and para-medical hands are available in required strength. He discounted chances of breakout of disease. Due precautions have been taken about quality of drinking water there, he said.

At least, 15 columns of the Army, including engineers and doctors, State Reserve Police and Central Industrial Security Force under the charge of Inspector General S.M. Shingari, 3,500 police trainees and 20 IAS probationers have spread out to provide relief and kitchens have been set up in every affected village.

People and organisations are contributing liberally for relief work. The administration today appealed to people not to contribute old clothes.

Mr. Raghunathan said that Government would, if necessary, stop the mushroom growth of "relief funds" being set up to collect contributions from public. He said, Government would prefer all contributions to be made to the Chief Ministers Relief Fund and cheque-payments would be preferred.

Couttesy : The Hindustan Times

5.2.5 ASSURANCE BY RAO ON RELIEF

New Delhi, Oct. 2 (PTI) — Prime Minister P.V. Narasimha Rao, tonight assured the people affected by the killer earthquake in Maharashtra that no effort would be spared to provide them relief and he appealed to the people to contribute liberally towards the rehabilitation work.

In a nationally televised appeal, Mr. Rao conveyed the nation's full sympathy to those in Maharashtra and Karnataka, hit by this "unprecedented" tragedy."

Thousands and thousands had lost their lives and the devastation was so vast that even now the full report of it had not been received, he said.

Whatever was humanly possible by the Centre, the State Governments and Non-governmental organisations was being done to rush relief. The Prime Minister said pointing out that the army had already been deployed and was doing a very good job.

Mr. Rao said that relief work taken on hand would be completed and no effort would be spared in this direction.

The Prime Minister said the task of rehabilitation was going to be stupendous and appealed to everyone to contribute liberally to the Prime Minister's relief fund.

Courtesy: The Hindustan Times

5.2.6 RESCUE WORK GOING ON IN FULL SWING

Harish Bhanot and Agencies

Bombay, Oct.3 — Nearly 95 per cent of the rescue operations in earthquake struck Omerga region and 90 per cent in Killari zone are over. Relief work is in full swing and rehabilitation phase will begin in a week's time, disclosed Mr. Raghunathan, Chief Secretary, Maharashtra.

Meanwhile, a mild tremor was experienced at Belkund village in Ausa Taluka of Latur district at 4.30 a.m. according to reports reaching the district collectorate earthquake cell in Latur.

The tremor lasted for a few minutes. It was of very mild intensity, an official spokesman said. The village had experienced tremors of moderate intensity yesterday also.

Mr. N. Raghunathan told journalists that the morale of the people in the quake-struck areas is high and they are showing a high degree of resilience. They have urged Chief Minister Sharad Pawar and the administration to rush seeds to them, so that they do not miss the next crops.

He said relief materials from abroad had started landing here today evening. A French consignment arrived at Sahar International Airport at 6.00 p.m. and the first US relief C-5 aircraft with 100 truck loads relief materials reached in evening and another aircraft, with a similar load of materials is slated to land a little after midnight. These aircrafts had made non-stop 21 hour flights from US bases to Sahar.

Mr. Raghunathan said the Hague based international organisations "Doctors Without Frontiers" is rushing doctors and medicines. The UK is rushing 80 tonnes of water-treatment tablets and other materials. A large number of countries including Pakistan, Japan and Israel had made enquiries as to what should they rush for relief and resettlement. Huge quantities of blankets, clothes, tents and plastic rolls for providing shelter to the people are on way to India and should be reaching within the next 24 hours.

The local people have overcome the initial shock. An excellent level of coordination among the various elements engaged in rescue and relief operations, led by Chief Minister Sharad Pawar from the front has helped the quake-struck people regain confidence.

Mr. Raghunathan reiterated that the number of dead in the Omerga region was 3,358 and the figure at Killari is 5,800. With rescue operations nearly completed, Mr. Raghunathan checked up with the Chief Minister on telephone in presence of this correspondent and said, the figure of loss of human life caused by the earthquake might go upto 10,000 at the worst. He added, the figure being reported beyond this was not factual.

He said there was no shortage of doctors, para-medical staff, medicines and bottled blood at the site. Medical centres had been set up in each one of the affected villages. When Mr. Raghunathan told Mr. Pawar, during the telephonic conversation that a journalist had brought to his notice, that the affected people were lying in the open compound of sugar mill at Killari, Mr. Pawar said that such information was wrong, as the people had been lodged in the large godowns of the sugar mill.

Mr. Raghunathan said, the people and organisations were sending relief materials and contributions to the Chief Minister's relief fund most liberally.

Meanwhile, the guardian Minister of Latur, Mr. Vilasrao Deshmukh said that the inflow of blood and medicines from various health centres was so much in excess that they were in the process of being returned.

In an appeal to people, he said what was needed at this moment was only items of daily needs.

Police has erected barricades, just outside the city limits for security reasons in view of the Prime Minister's visit tomorrow and also to check the flow of onlookers into the disaster sites.

Rescue teams battling rain and other odds intensified effort to clear the bodies and being succour to the survivors of Thursday's earthquake in Maharashtra, in which the death tool was officially put at 21,500 and it likely to scale upto 30,000 report agencies.

Even as the unprecedendted peacetime relief and rescue mission progressed in the quake-hit villages, Balakund village in Latur district, one of the worst-affected experienced tremors last night in which three persons were feared killed after their houses were flattened.

Rains impeded relief work last night and adding to the woes of troops and others engaged in the operation, created puddles and slush, hastening the decomposition of the bodies and carcasses of livestock.

Relief materials are pouring in generously, but have thrown up several problems in its wake — like storing them and the authorities turned away some bands of volunteers as collection of too many of them could obstruct relief work.

A Vayudoot Avro aircraft this afternoon airlifted three tonnes of food and medicines to the survivors of the disaster.

SURVIVORS' TALES: Survivors had only tales of woe to tell newsmen from Gulbarga, who visited some of the affected villages, including Peth, Sangavi, Sastur, Rajegaon, Nandurga, Mangrul, Yekkundi, Holli, Tavasi, Rebi Chincholi, Koti, Tani, Limbala, Gubul, Hiregaon, Kankigod and Kaoutam.

Almost everywhere the victims are left to fend for themselves with the relief material yet to reach many of the affected villages.

Grim-faced villagers rummaging through heaps of boulders and mud dunes to salvage whatever remained in their once magnificent houses are a common sight.

The men in olive-green and policemen have nothing much to do as the villages carry household articles and tin sheets retrieved from the ruined houses to makeshift tents, they themselves had erected without waiting for the Government to come to their aid.

No semblance of any salvage operation was seen in most of the villages even three days after the tragedy struck. "The help from the Government we got was wood to light the mass funeral pyres and a few bedsheets," lamented V.N. Bhonsale of Peth Sangavi.

Courtesy: The Hindustan Times

5.2.7 MASSIVE AID OFFERS FROM COUNTRIES

New Delhi, Oct. 3—The government has received offers of massive aid in cash and kind from several countries for the earthquake victims of Maharashtra.

Officials sources said the Crisis Management Group of the Government was monitoring relief work in the affected areas on a daily basis and meeting took place today under the chairmanship of Relief Commissioner K.B. Saxena.

The American Government has sent aid worth over US $ 10 lakh, while France has given about US $ 92,000, Italy US $ 312,000, Japan US $ 25,000, Norway US 1,90,000, UK US $ 5,40,000 Canda US $ 2,25,000, Switzerland US $ 1,45,000. The Department of Humanitarian Aid of the UNICEF has given US $ 50,000 and CARE has given US $ 2,50,000, the sources said.

The Maharashtra Government has informed the Centre that all material be sent to Bombay as it would be easier for the aid to be sent to the victims from there, the sources said.

Among the priority items listed by the State Government are prefabricated housing material, water purification tablets, vegetarian food and medical equipment.

In its attempt to speed up relief operations the Government has been exploring the possibility of obtaining sophisticated equipment to detect survivors trapped in the debris and it has received several offers from abroad.

A high powered team led by Minister of State for Defence Mallikarjun and including Air Chief Marshal S.K. Kaul and General B.C. Joshi today visited the earthquake hit areas. The team flew in an IAF Boeing and then visited the area in an MI-8 helicopter.

Briefing the media the Air Chief Marshal said the IAF had deployed seven AN -32 aircraft and MI-8 helicopters (Chetak) from September 30 till, date. The Air Force has delivered over 55 tonnes of medicines and other essential supplies flying from Delhi to Hakimpet, Yelahanka, Bidar and Pune. He congratulated the pilots involved in the round-the-clock supplies in spite of bad weather and poor visibility.

Relief contingents of the Indian Navy from the Eastern as well as the Western Naval Command have started functioning from Solapur and are undertaking relief work in the worst affected Umarga-Killari region.

The Joint Naval medical team consisting 14 officers and 46 assistants is carrying about one tonne of essential drugs and medicines. A logistic support team deployed there is carrying blankets, durries, rice, dal, sugar, milk, powder, bread and other relief materials. So far the teams have attended to over 800 stricken people.

A large number of non-governmental organisations have also sent material for the earthquake victims with Australia despatching a planeload of nutrition biscuits, while water tablets have been sent by the British Government, the sources said.

The Railways have set up two galvanised iron sheet shelters for the affected families in the area and is setting up more such shelters, the sources said.

The Petroleum Ministry has said there was no difficulty in supplying kerosene and diesal and any other material that the State Government may want.

The Ministry of Surface Transport has said that all the major roads and highways in the area have not been damaged by the earthquake and therefore the lines of transportation could be kept open, the sources said.

Courtesy: The Hindustan Times

5.2.8 CONTROVERSY OVER LAST RITES

Latur Oct. 3, (HTC) — Even as the Assam Regiment are busy in digging out the bloated and decomposed bodies from the mounds of rubble and consigning these to fire, a controversy has broken over the funerals. Family members and relatives of the deceased have strongly opposed the mass herding of the bodies and consigning them to flames without any rituals.

Even the use of kerosene and petrol to light the pyre is disapproved by the relatives. They feel that the deceased wouldn't attain 'heaven', unless the necessary ritues as per "shastras" were following and the bodies were consecrated by the priest before being consigned to the flames. In a couple of cases, the relatives carried away the bodies to the villages of distant relatives so as to perform a "decent" funeral.

In Sastur village, which has a sizeable Muslim population, many survivors objected to the mass burial of the dead.

Courtesy :The Hindustan Times

5.2.9 RS. 50 CR. AID ANNOUNCED

No efforts will be spared for rehabilitation, says PM

Killari, Oct. 4 (PTI, UNI) — Prime Minister P.V. Narasimha Rao, today announced an immediate Central aid of Rs. 50 crore and promised all help to the survivors to begin lives afresh after he surveyed the earthquake-devastated villages in Osmanabad-Latur belt of Marathwada region of Maharashtra.

"Money was no constraint and we will spare no efforts for rehabilitation, a visibly shaken Mr. Rao said, as he moved from village to village and heard the traumatised survivors during his mission to provide a healing touch.

The Government would embark upon a massive programme of seismological mapping of sensitive areas of the country to avert the Killari like disaster", Mr. Rao told newsmen after the visit, which took him to hamlets virtually converted into graveyards by the nation's worst-ever quake that has left about 30,000 people dead and 12,000 injured.

He said: "It is very essential to start such an exercise as the earthquake has shattered some of our beliefs and conceptions regarding quakes."

At Sastur in the worst-hit Omerga district the Prime Minister saw mountains of rubble, bearing mute testimony to the tragedy that befell the villagers after the quake reduced their dwelling to ruins, leaving little for the survivors to live for.

Mr. Rao consoled Kashinath Kumbar, whose hopes and drams lay buried in ruins having lost seven members of his family in the predawn knock of death by the quake.

"I have visited many quake affected areas, including Uttarkashi but this earthquake is the worst one," Mr. Rao said sharing his assessment of the grim situation with reports at Latur after his visits to some villages and an aerial survey.

He said the immediate priority would be to ensure that the villages devastated by the quake were rebuilt at safer sites, a plea the villagers also made to him, and the work would commence soon.

Even as more and more foreign aoid flowed from governments and voluntary agencies, thousands of troops and armies of volunteers galvanised their efforts to bring relief to the survivors braving all odds.

Tin sheds, tents and other temporary shelters have been set up in 30 quake-ravaged villages in Osmanabad district and nearly 10,000 survivors were being fed twice daily.

Teams of doctors, dawn from the Army and Navy and Government were dispensing medical aid, aided by bands of para-medical volunteers.

Armymen continued to scramble through the piles of debris continuing their hunt for the dead for the fifth consecutive day today, even as the threat of an outbreak of epidemics loomed large.

With mass funeral continuing funeral pyres glowed in several villages, mass burials were also a common sight.

Curious onlookers streamed in a steady flow to the village, as police struggled to keep them at bay by cordoning off the area. Hundreds of trucks and other vehicles also came, causing traffic snarls.

Addressing small roadside gatherings of villagers, Mr. Rao said, I know your loss is great and irreparable, I promise you maximum help.

He said the toll now stood the tool at 10,000 but it would not be possible to give a final figure until the full areas was cleared of the rubble.

The Prime Minister, accompanies by Lok Sabha Speaker Shivraj Patil, Maharashtra Chief Minister Sharad Pawar and leader of the Opposition in the Lok Sabha Atal Behari Vajpayee, landed in Bidar in Karnataka by a special aircraft. From there he took on a helicopter to make a survey.

Courtesy : The Hindustan Times

5.2.10 DIGGING OF BODIES GAIN MOMENTUM

Sastur (Osmanabad), Oct. 4 — The operation to dig out bodies lying under tonnes of stone and mud gathered momentum today with the arrival of sniffer dogs and special equipments. Several dogs of the Maharashtra police arrived here early morning and were pressed into detection job in some of badly affected villages like Killari, Sastur, Talini, Kavata and Ekaundi. A team of French disaster management experts also arrived here from Bombay this morning to aid and advice the administration in rescue operations. The team has also brought a sniffer dog with them.

Meanwhile, local authorities at Omega said they were expecting a special equipment from the Defence Research and Development Organisation (DRDO) which can help in locating bodies lying deep under the debris. Till now the armed personnel were conducting operation primarily on the basis of smell to locate the bodies and excavate them. The instruments is claimed to be a portable one and could distinguish the human body from animal carcasses.

Meanwhile, the people in some of the affected vilages spent their fourth consecutive day under the open sky. A light drizzel last night also added to their woes as they did not have any place except few trees to take shelter.

Though the Government had announced that creation of temporary shelter was one of the topmost priorities, in actual practice seems to have received low priority. After waiting in vain for four days, the villagers were seen erecting temporary sheds in the fields close to villages by retrieving bamboo poles, wood beams, and whatever of use from their collapsed houses. Maharashtra Minister for Rural Development Ashok Chavan said a sizeable amount of roofing material like galvanised sheets and tents had reached. But these seemed less than adequate. The sitution is expected to ease only on Wednesday when about 50 trucks of roofing material donated by various organisations is expected to arrive.

Courtesy: The Hindustan Times

5.2.11 EPIDEMICS THREAT FEARED

Killari (Maharashtra), Oct. 4, (PTI) — Areas devastated by Thursday's quake now faced the treat of epidemics because of rotting bodies, chocked drains, lack of clean water while Army and civilian teams today continued with rescue and relief operations in Latur and Osmanabad districts of Maharashtra.

Medical teams from the Western Naval Command which warned of the epidemic spoke of the strong stench of bodies rotting in mounds of debris which is all that is left of 32 villages in Osmanabad district. The navy teams are treating survivors at the rural health centres in Omerga and mobile field offices.

Meanwhile, civilian rescue teams and the armed forces were working on a war footing to provide relief while generous aid from local and foreign agencies poured in for the quake victims. Tin-sheds tents and temporary shelters have been erected in the affected areas and makeshift hospitals and community kitchens set up.

Many voluntary organisations have also adopted ill-fated villages and were providing food and clothing to the people. A report from Omerga said telecommunication links and power have been restored in many villages of Osmanabad district.

Despite intermittent rain, Armymen are sifting through mounds of debris to search for more bodies.

Courtesy: The Hindustan Times

5.2.[illegible]2 LATUR ROCKED AGAIN

From Anil Anand

Osmanabad, Oct. 5—The areas bording Andhra Pradesh including Latur and those hit by the tremor last week, were once again rocked this morning, further creating scare among the people. The shock measured 4 on the Richter scale but there was no damages.

As the debris in most of the affected villages were still to be cleared, effort to locate more bodies buried under the collapsed houses took another turn with the arrival of a French team bringing trained dogs. The 10-member team belonging to a non-government organisation named Medilor, has brought four dogs. The canines have a history of saving a number of lives in the earthquake hit areas in different parts of the world since 1989. Apart from the sniffers, the team is also carrying sensitive equipments which record the movements of the buried survivors and amplifies it to be heard outside.

A daunting task waiting the authorities ahead was to take care of the large number of orphaned children. The process to identify them, has been started. According to a district official, uptill now 400 such children have been found. The children left to fend for themselves at this tender age, will require surrogate parents for rehabilitation. However, no voluntary organisation or individuals have so far come forward to lend them a helping hand. The only offer made so far was from the SOS villages.

Many such children are already undergoing treatment in various hospitals. Till now, they are under the care of the medical staff. But event they were not sure about the future of these children.

There is another section of the victims, those at the fag end of their lives, left without a caretaker, still to recover from the shock of having lost their entire family, have almost become numb.

Negaba of Holi village is one such victim. At 80 he was sitting alone on the rubble of his small hutment where he lived with his five family members. He had no answer about his future plans. Similar was the plight of Pancholi, 65, of village Bajewadi. She burst into tears and wailed loudly when some relatives came to meet her, three days after the tragedy had struck. Maruti Jadhav left alone narrated his future plans. I will dispose of my land to clear debris and wait for the last day," he said.

Fully knowing that surviving alone will be tough at this age, they were cursing themselves to have survived the catastrophe while their children perished in front of their eyes. Nagaba who had gone to Pune on the fateful day, was disgusted with himself that why he was out while his family perished. Baba of Peth Sanghvi survived because he had gone to the fields early morning while his five family members including sons, wife and daughter in law slept inside the house.

Most of these persons looked a mental wreck. They have become stoic and hardly few reacted to queries. They simply stared with their silence and blank looks conveying the feelings.

BOMBAY: Maharashtra Chief Minister Sharad Pawar today calculated that the rehabilitation of the earthquake struck in Maharashtra would cost Rs. 900 crore reports HTC.

He added the rehabilitation of the quake-struck, including the construction of new quake-resistent houses and townships would be started on the Dussehra day 24 next.

Mr. Pawar who was talking to journalists here, on his return from the quake-struck areas said the work of rubble-rummaging and extraction of bodies from under the debris would be completed in the next two days. The Army was finding only five bodies a day at present.

He said the death toll, till today late evening was 9,774 and this may go up by a few more. But, it appeared clear that the total would not exceed 10,000. Mr. Pawar, who had landed by air at the earthquake hit area within three hours of the quake and had led the rescue and relief from the front all these days agreed that it was a major disaster and destruction was extensive but there was practically no chance for the figure dead going beyond 10,000. He added the death toll had been thoroughly verified from records, survivors and other know ledgeable sources like village officials.

A correspondent asked Mr. Pawar, "God forbid" if earthquake of the intensity and extent of Omerga-Khillari areas hits megapolis Bombay, are you prepared to face the situation."

Deeply upset with the question Mr. Pawar told the questioner: "may there be a lot of salt in your mouth". Another correspondent added "lot of pungent chillies too."

Courtesy: The Hindustan Times

5.2.13 NO WATER TO QUENCH QUAKE VICTIMS THIRST

Omerga, Oct. 5 — The worst is yet not over for the survivors of the earthquake which rocked Osmanabad and Latur districts of Maharashtra. The much-talked about operation is yet to move in a planned manner despite aid being received from all parts of the world. The dazed survivors, particularly in the partly affected villages, were still bereft of an essentials like safe drinking water.

Even among the survivors, lucky are those living in villages like Killari, where the Army has taken charge of the relief operations. Half-a-dozen other villagers which this reporter visited today, were still to attract proper attention of the authorities, for instance at Nimbala, Talani, Hublal, Rejewadi and Ekodi, both medical and protected drinking water facilities were scant. They were still drinking water from the wells which were untreated and could cruise serious health hazards in the days to come.

A glaring instance of the missing coordination among the various Government agencies, is the improper utilisation of the number of doctors who have come from all parts of the State as well as the neighbouring ones. As a result, many injured persons in these villages have engaged themselves in slavaging their belongings from the debris without even getting their wounds treated. Perhaps they have neither the means to reach the district hospitals for the time as their first priority ws to extricate was left of their meagre belongings.

The innoculation drive, which should have been launched immediately after the extrication of bodies was still confined to district headquarters though in Killari one of the worst-affected villages, the Army doctors were already on the job. The only measures taken by the state Government agencies to prevent infectious diseases was spraying bleaching powder. This was also confined to villages whose names have prominently figured in the Press.

These drawbacks exist despite the presence of about 10,000 doctors and other para-medical staff belonging to the health departments of two affected districts and hundreds other who have come from outside. Member of a medical team, which has come from Bombay, said that they were still groping in the dark not knowing what to do, as the local authorities were not coming forward to assign task to them.

Courtesy: The Hindustan Times

5.2.14 RESCUE WORK CONTINUES

Killari, Oct.7 (PTI) — Amid fast-paced efforts by authorities to revive normalcy from the ruins, rescuers continued to uncover decaying mutilated bodies even seven days after the collosal earthquake devastated villages in Osmanabad-Latur belt of Marathwada region of Maharashtra.

Reports said as the shell-shocked survivors were coming out of trauma, consignments of seeds of jowar, safflower and sunflower, major crops of the quake-hit belt have been rushed to villages in the worst-affected Latur district.

While the relief and rescue mission was going at full blast Armymen have been asked to erect by Monday, tents as makeshift the schools in several villages where the killer quake left roughly 4000 children orphaned.

State authorities said the people were demanding seeds as ideal moisture and soil conditions prevailed. But, the holocaust survivors in some villages, mired in the tragedy said they were in no mood to resume the sowing operation.

The over-powering stench of decaying flesh filled the air, as troops and volunteers went ahead with their daunting task of removing the bodies from heaps of debris still lying across several villages.

Meanwhile, the Central team comprising experts on housing health and science and technology visited some affected villages to assess the havoc wrought by the quake.

The Railway has rushed a special goods train carrying 24,000 cartons of imported edible oil and 8,000 bags of wheat from Wadibandar Goods Depot of Bombay division of Central Railway to Solapur free of charge as a special gesture.

Maharashtra Chief Minister Sharad Pawar conducted surprise visits to some villages yesterday apparently to ensure the relief mission was smooth and the materials reached the needy.

Meanwhile the Government of the Netherlands has allocated three million Dutch Guilders (Rs. 5 crore) for aid to the victims of the earthquake that hit Maharashtra, Andhra Pradesh and Karnataka.

A plane load of medicines and foodstuffs the Kuwaiti Government flew into Bombay for despatch to the victims.

HYDRABAD: The recurrence of the earthquake in the city has a very less probability, according to NGRI scientist B.K. Rastogi.

The aftershocks of the quake is usual phenomenon he said adding that another quake with the same intensity has no scientific support.

He refused claims made by astrologers that another quake would hit the city in the second week of October saying the forecast of quakes has no scientific basis.

Courtesy: The Hindustan Times

5.3 DEATH WRITES AN EPITAPH

The spiralling smoke and embers from the multitude of pyres dotting the rain-drenched landscape give an eerie feeling of helplessness. There is tragedy beyond compare.

An army of sombre-looking men and women are at work, for reviving life in the cluster of once throbbing villages that were, in 'quaking' seconds, turned into a graveyard.

The glimmer of hope in uttar hopelessness in what looks like a no man's land, writes Deepa Gahlot, sparkles in the dedication and selflessness of the survivors and volunteers who braving the rain, the rubble and fear of epidemics and fresh tremors are attempting to counter nature's wrath.

Shivajirao Bhagwan Waghmode has a small timber mart in Solapur. As soon as he heard of the earthquake that had devastated villages in neighbouring Latur and Osmanabad districts, the first thing that struck him was: who will cremate the dead? A question which could have occurred only to a deeply religious and unselfish person. He rushed to help the only way he could. He loaded two trucks with firewood worth Rs. 5,000 and rushed to help extricate bodies and cremate them. He performed this simple but extraordinary deed without hope of any reward, not even his picture in the papers. This is the dignified face of charity.

There are also those who came to help and stayed to strip the dead. There are the fly-by-night opportunists stalking the streets from Bombay to Latur with collection boxes, hoping to tap the lazy consciences of people who can't or won't spare more than a few coins to assuage their conscience. There are the city slickers who come with their pious expressions and self-righteous attitudes to "help these unfortunate people." This is the irksome side of altruism.

Whenever there is a catastrophe, there are four kinds of people you find: victims, volunteers, voyeurs and vultures.

The scene at Killari village is horrifying — this is an understatement. No amount of packaged torment on TV can truly capture the scene.

In spite of the police cordon, there are hords of people who have come to gawk. The victims have been reduced to performing animals and seem to be willingly playing the role. Maybe it is shock, maybe an unnamed emotion that is experienced after the mind has gone beyond anguish.

A young tailor in Killari takes me to his devastated home, and with a voice that is, surprisingly, tinged with pride tells you how many of his relatives died. There is a strange exhibitionism about the people who gather around to relate tales of suffering and loss. Like a leper who displays his wounds to get alms. Maybe they just want to unburden themselves to anybody who will listen. At the hospital in Solapur, a woman sees a camera and whips back the sheet that a covering her child's broken limbs. What made her subject her child to be curious gaze of the camera and the callously probing questions of strangers ?

Everywhere there are survivors who impassively, and with a kind of solemn ceremony, pick up whatever is left of their homes. The wood from the crushed dwellings has already been used to cremate the dead. In the background are pyres burning and the air is thick with the stench of rotting flesh. But they are oblivious to it. They pack up useless items. One man rescues his audio-cassettes from the rubble, while a women frantically digs with a sickle to salvage out her dented cooking vessels. Another picks up tin sheets to make a make-shift shelter in the fields. Little items like a string of plastic flowers and a powder tin indicate that people lived happily in houses of wood and stone. Now they do not want to return to their once prosperous village. "What do they go back to?" says a former sarpanch of Killari. "Now the situation is that whoever you meet is alive, the rest are all buried in there. This is graveyard". Papers of accounts ledgers from the Bank strewn around in the slush also suggest a dead-end for the survivors. They have to build their lives all over again.

One thing that strikes me is that people who have lost everything, express no grief, shed no tears. They are either cold with shock or have amazing fortitude. The acceptance of pain seems to be dinned into the smallest child. A skinny little kid in the hospital is having his ugly stinking wound cleaned rather roughly by a ward boy. But he bears excruciating pain with not more than soft whimpers. A villager says that people of these parts do not weep. "You city people weep and forget in ten minutes. These people will keep all the pain bottled in for years and probably cease to be men."

"We have no emotions left" says a young survivor. "We don't even have the courage to pull out our dead. At least we are alive to tell our tale, there are homes in which there is nobody even to mourn the dead."

Every time there is a patient listener, there is a clamour of complaints and demands. "Nobody has come to help us," the survivors complain. "We need food, water, shelter, money to rebuild our lives. Nobody is even willing to listen to us. They all come up to the outskirts and go away. The policemen who were supposed to help us have stolen our property. Who shall we turn to?"

As bulldozers flatten their dreams and helicopters circle overhead, they stand and stare. Once in a while a curse escapes the lips of one of the men.

On the remains of his shattered home an old man sits disconsolately and waves away sympathisers. He has lost all his family and wants to bear his grief alone. He has been sitting there, in the same position for days, turned to stone.

A dog weeps for its dead master.

A very efficient-looking army camp is set up on the road to Killari. Relief workers from the army, reserve police and government organisations are working round the clock, taking the wounded to hospital, trying to reach food and basic necessities to the victims but whatever they do doesn't seem to be enough. A bureaucrat admits that so far, most of the relief material has come not from the government but from private donors.

It is amazing how many people have just dropped whatever they were doing and rushed to do good work. For instance a small team of amateur mountaineers have volunteered to dig out bodies. Hundreds of vehicles in all shapes and sizes loaded with people bringing food, medicines and cloths, with banners proudly proclaiming the names of their organisations — from autorickshaw unions to garment traders, organisations, tyre dealers, associations and other such odd, outfits to religious groups and political party workers — they are all there.

A lot of their efforts are useless. They are indignant at being turned away by the security men. Food they have brought has decayed. I wonder if they really care or are they just doing it so that they can tell their friends and neighbours that they did their bit for quake victims. And at the next tea party or cocktail gathering, tell horror stories to task-taking listeners. Would these people who have driven long distances and spent thousands of rupees, hike their servants, salaries by a few rupees or stop rejoicing everytime a slum colony is torn down ? Anyway they are well-meaning and probably wanting to prove to themselves that they have a heart. But their insistence on handing out things to victims themselves does nothing but cause traffic jams.

In the hospitals doctors are doing an admirable job. But these days nobody believes in selfless service. They want to make sure that their effort are recognised. They want to let it be known that they are doing great work coping with the emergency. That is only natural — everybody wants recognition of their successful missions. Hundreds of fractured limbs have been put into plaster, emergency operations performed. They have also tried to organise psychiatric care for those in a state of trauma.

Thousands of people have been making trips to Killari and neighbouring villages to see the 'sight', their noses covered with handkerchiefs. They have reduced the tragedy to a *tamasha.*

They have been able to dodge security and rushed in through the bylanes and fields. They come with curiosity and complacency on their faces. Their colourful plastic water bottles slung across shoulder, picnic lunches packed and their little camera clicking away for the albums back home. They want to grimace at the bloated and decomposed bodies and screw their noses at the smell. They want to test their endurance and feel how brave they are to be able to stand what they see.

A man has come with a child to "show him around". Soon there probably will be tea-shops, pan-and-cigarette stalls, bangle-sellers and merry-go-rounds in the desolate villages.

I honestly cannot say that all the journalists who are converging at the quake sight — mainly the easily accessible villages like Killari and Satur

— from all over their world, with their hitech equipment and seen-it-all scowls are not voyeurs too. We are all looking for photo and story opportunities.

One man's misery is another man's merchandise.

"We are stealing from them, is it?" sneers a policeman, reacting to the accusation that they came to help extricate bodies and stole whatever they could find. "We are also living in camps here. Where will we hide the booty?"

But many of the people who rushed to the quake-affected villages after the quake havoc, robbed the dead and dying. In another villages, a bunch of Lamadi tribals — not rendered homeless after the quake, because they are nomads, are feasting on the food meant for the hungry and the sick.

The clothes collected by volunteers are being grapped by people who clearly don't need them. An old man in a *dhoti* pounces on a two pairs of trousers as a little girl finds a colourful skirt in a pile of clothes and whoops with joy. A woman who has managed to collect a huge pile of clothes examines the loot with joy.

But the worst are the 'leaders' who have come to make political capital of the calamity. The Shiv Sena with their orange flags and scarves and RSS men in khaki shorts are hard at work and according to villagers, they are helping. But a lot of *khadi* topi and safari suit types are rushing around to no purpose but to be seen at the right place at the right time. Minor party functionaries are flexing their muscles all over the place. A congressman says indignantly that the others are flaunting their insignia. While their party men are doing the real work. A Janata Dal man accompanying a Buddhist member of the Minorities Commission cribs about lack of coordination, but does nothing to help lessen the chaos, ease the pain. It seems everybody is feeding on the disaster, ripping the flesh off the situation like a bird of prey rips the meat off its victim.

Already the tragedy is being relegated to the inside pages of the newspapers. In a few months people will have forgotten all about it. But that is when real help will be needed, to heal the minds of the survivors and help them return to the land of living. Does today's shocked and concerned world have that much time and patience ?

Courtesy: The Hindustan Times

5.4 PREDICTION OF ANOTHER EARTHQUAKE IN MAHARASTHRA

5.4.1. EXPERTS RULE OUT ANOTHER TREMOR

Hyderabad, Sept. 30 (UNI) — Seismological experts have ruled out the possibility of another earthquake in the near future in the Latur region, which was hit by a powerful tremor this morning.

According to National Geophysical Research Institute (NGRI) Director Harsh K. Gupta, this morning earthquake was followed by a few aftershocks — four of them were recorded distinctly by the NGRI Seismological Observatory here.

He said one was felt deadly after 45 minutes and had a magnitude of about 4.5 on the Richter Scale.

He said there were several queries about the relation of this earthquake to reservoirs and clarified that it had nothing to do with the Koyna dam or any other reservoir.

The Whispering gallery of the world famous Golgumbas in Bijapur district in Karnataka developed cracks in the earthquake that shook parts of the State this morning.

According to information reaching here, the outside walls of the over 300-year-old tomb have also developed cracks.

Panic gripped the citizens of the twin cities as rumours of the recurrence of the earthquake that devastated the Marathwada region of Maharashtra this morning spreads like wildfire.

The city, which experienced the tremors distinctly woke up to a normal day, after the initial shock. But, no sooner did the people realise the devastating impact the earthquake had caused in neighbouring Maharashtra, fears surfaced with the rumour mill working overtime.

In terms of magnitude, this is the biggest tremor to be felt in the State in recent times with the National Geophysical Research Institute here recording 5.5 on the Richter Scale, 0.1 more than the one recorded on April 13, 1969 at the temple town of Bhadrachalam which read 5.4 on the Richter Scale.

The earthquake, which claimed thousands of lives in southern Maharashtra today, was like the 1967 Koyna earthquake caused by the "rubbing together" of landmasses on the two sides of the 400 km-long Jurdavadi rift, spreading from south-east of Sholapur and ending in the north of Pune, according to eminent geophysicist Dr. J.G. Negi.

The Kurdawadi rift, having a width of between 40 and 60 km, was the actual epicentre of today's earthquake.

Courtesy: The Hindustan Times

5.4.2 ANOTHER EARTHQUAKE IN DECEMBER?

Nagpur, Oct. 2 (PTI) — Another earthquake is likely to hit Maharashtra by coming December-January, according to Dr. P.C. Adyalakar, former Director of Geological Survey of India (GSI) and Central Groundwater Board.

The tremor may affect areas east of Western Ghats as the area has hydraulic continuity with Koyna through lineaments and, in effect the area of the Sept. 29-30 earthquake was interlinked with the Koyna-Krishna region through a fabric of North-South and other directional lineaments. Lineaments are systems of underground fractures and fissures.

Dr. Adyalakar says the Latur area earthquake is linked to the 1967 Koyna earthquake as any medium-intensity earthquake measuring intensity earthquake measuring upto 5 points on Richter scale, with epicentre in the vicinity of Koyna was always felt in the Latur, Omerga-Killari area.

The lineaments activated in the present earthquake must be in a major zone of structural weakness in the form of faults, fissures and thrust areas, the geologist says.

Dr. Adyalakar says the criss-cross currents of rivers like Ghod Bhima, Sina, Maan, Manjra and others affect the Koyna lineaments in general, there being abundance of ground water in the Killari-Omerga-Krishna area. It had a tendency to go deeper and deeper through successive interflow zones.

In the case of Koyna the inter-flow zone had gone down to 33 Km, as can be inferred from interpretation of the inter-flow zones reservoirs, he says.

In the Deccan trap there are about 300 lava flows in the basaltic formations with an aggregate thickness of 3,000 metres and the basalts have developed potential medium grade squifiers deep underground water channels which may lead to earthquakes, Dr. Adyalkar says.

Courtesy: The Hindustan Times

5.4.3 SEVERE QUAKE PREDICTED IN BOMBAY

Hyderabad, Oct. 3 (PTI) — Bombay is sitting on a "satellite faults" and will suffer a severe earthquake, according to the same two scientists at the National Geophysical Research Institute (NGRI) here who had cautioned about the recent Maharashtra Earthquake 20 years ago.

According to Dr. N.J.G. Negi and Dr. Krishnabrahaman, Bombay is situated in a region "where the gravity gradient is the steepest and the gravity anomaly the highest", indicating that the region is a fault zone.

Dr. Negi, who is currently heading the Madhya Pradesh State Science Council, told PTI that occurrence of an earthquake in Bombay was certain "but nobody can predict when this will happen."

Dr. Nagi and Dr. Krishnabrahaman have been using the gravity data to locate faults under the surface that can become active and produce an earthquake.

The Maharashtra quake is believed to be due to reactivation of the Kurudwadi fault whose existence was first discovered by the two NGRI scientists in 1973 on the basis of the gravity data interpretation.

Gravitational force is not the same everywhere on the surface and sudden variations called "gravity anomalies" are suggestive of cracks or faults on the earth's crust.

A more recent publication by the two scientists said that the gravity anomaly over the Bombay region "is the highest" in India. According to the Scientists, Bombay lies over a "satellite faults" that can get active.

Courtesy: The Hindustan Times

6
Impact Assessment

6.1 IMPACT OF EARTHQUAKE ON THE ECOLOGY AND ENVIRONMENT

Every ecosystem on the earth has few basic components, which are linked together by a food chain, if any one from food chain get affected, certainly it will affect the others of food chain. In this regard the effect of earthquake could be well explained by the Fig. 6.1.1.

TERRESTRIAL ECOLOGY

Felling of trees decreases the water absorbing capacity of the soil and without vegetation soil particles become loose, resulting in the formation of water channels during monsoon, which ultimately induce the soil erosion and surface run-off, again a threat to the remaining vegetation and wild life.

AQUATIC ECOLOGY

After an earthquake, increased quantity of silt and sand particles in water, due to the run-off, affects aquatic animals in various ways, which ultimately results in the water pollution. Figure 6.1.2 and 6.1.3 are explaining the effect of earthquake on water bodies.

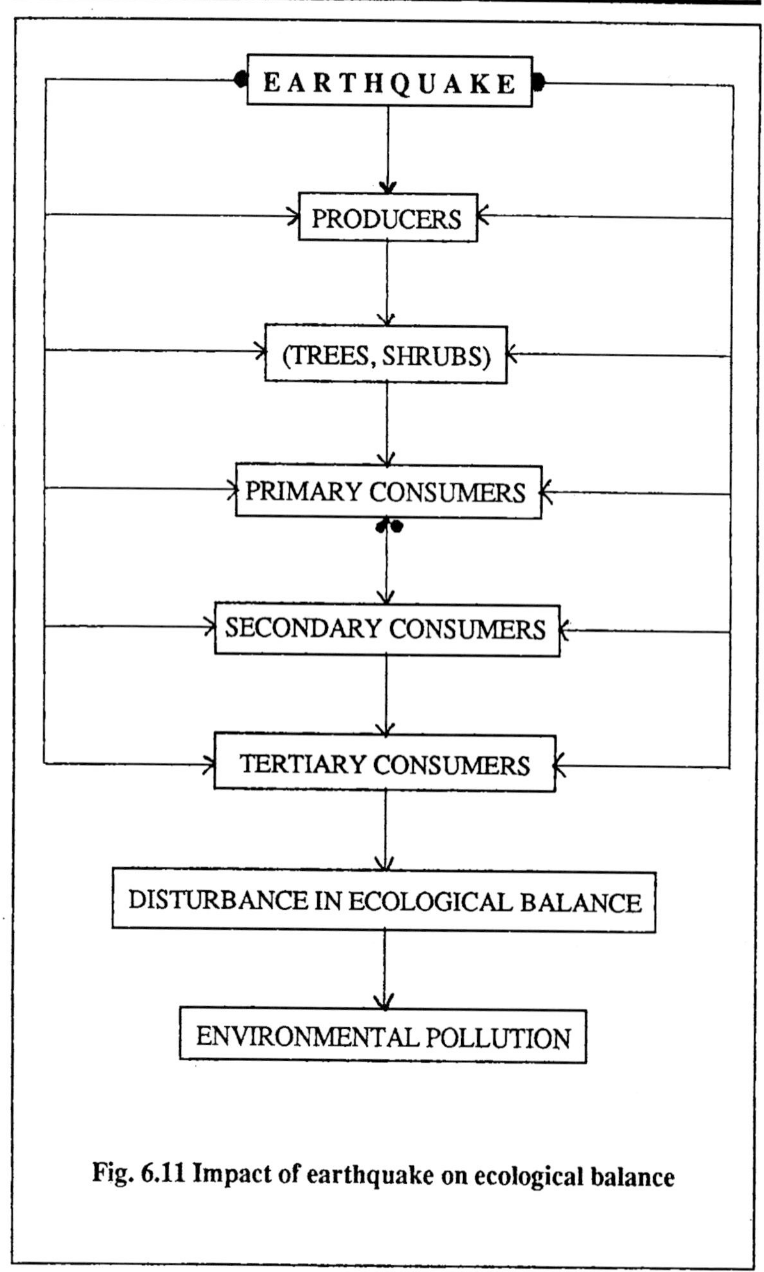

Fig. 6.11 Impact of earthquake on ecological balance

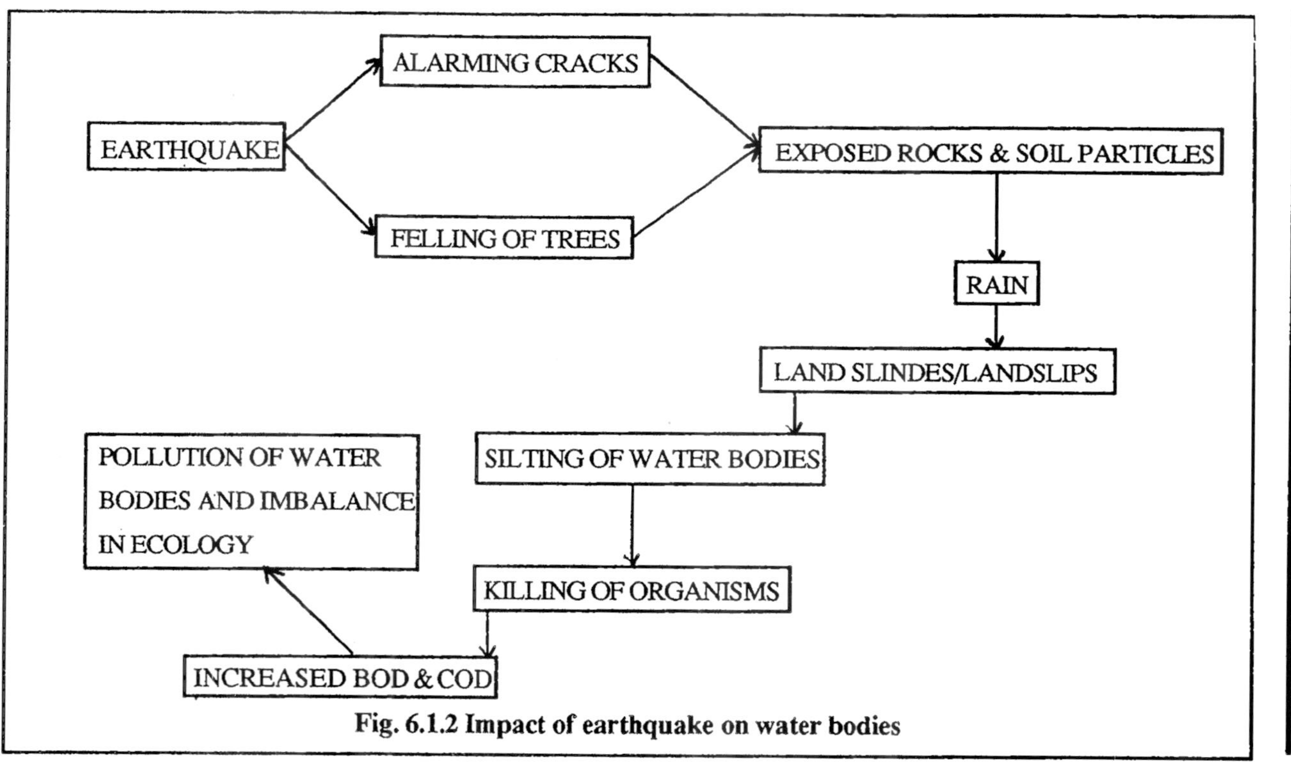

Fig. 6.1.2 Impact of earthquake on water bodies

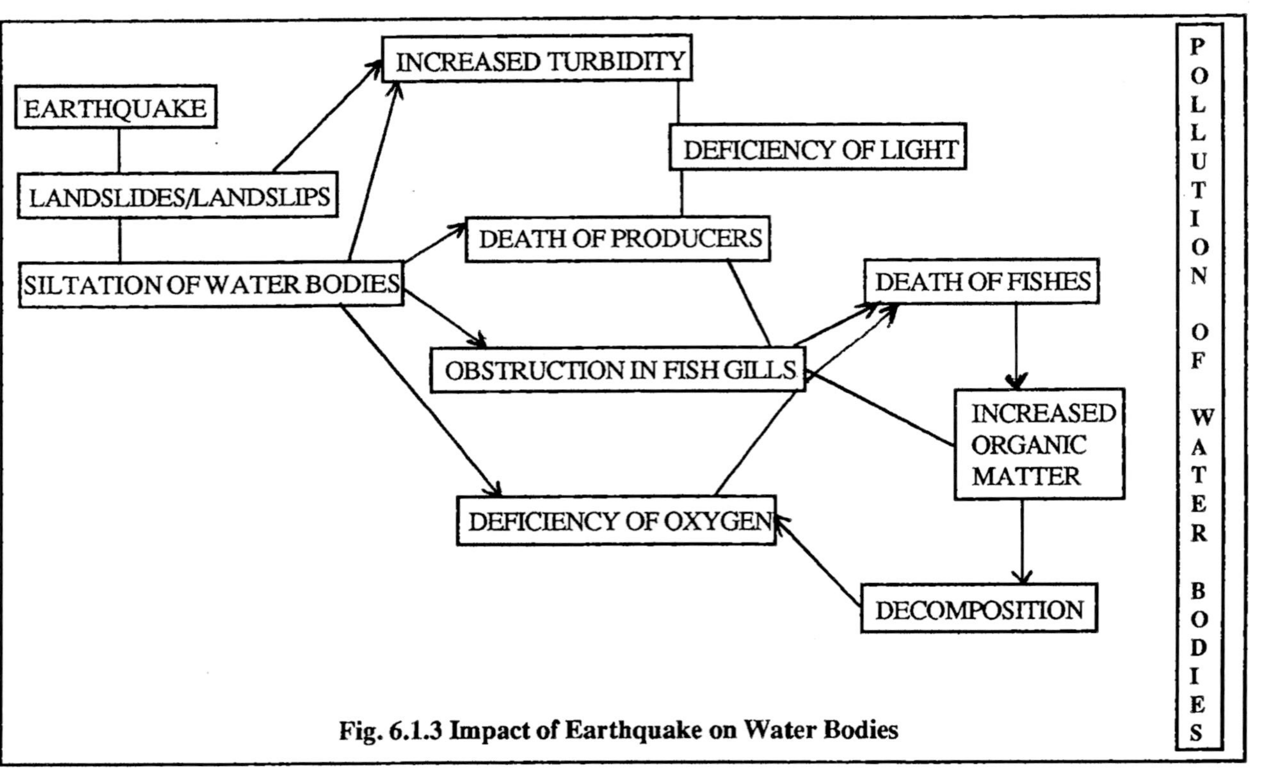

Fig. 6.1.3 Impact of Earthquake on Water Bodies

6.2 IMPACT OF EARTHQUAKE ON EDUCATION SYSTEM

An earthquake could affect the education system adversely. In a layman language there are no visible signs of damage, could be imparted by an earthquake, on the education system. But, if we consider all the conditions prevailed in the region of earthquake, one can understand easily the impacts of earthquake.

Various factors which could affect education system are — damaged school buildings, lack of shelters, clothes, food, damaged approach, routes, affected teachers and affected students. If we analyse these factors combinely then we could understand easily how an earthquake could create a problem to education system (Fig. 6.2.1).

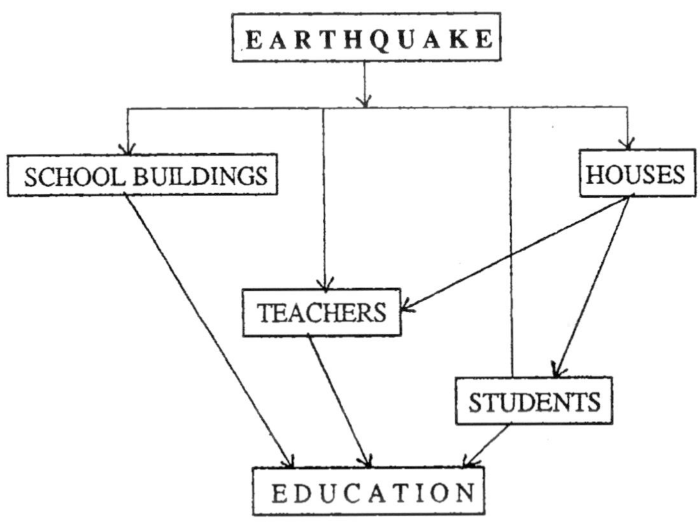

Fig. 6.2.1 Impact of Earthquake on education

6.3 IMPACT OF EARTHQUAKE ON AGRICULTURE

For taking a better crop yield, it is necessary to give all the basic requirements to the crop. Earthquake could cause damage to the fields, animals and working population. Basic chain of necessary things to get better yield is — good fields —> ploughing equipments —> animals or vehicles for ploughing —> seed fertilizer —> skilled and unskilled labour.

An earthquake could affect all these adversely. Which in turn could affect the agricultural yield for one or more seasons, it could be understand easily by the perusal of Fig. 6.3.1.

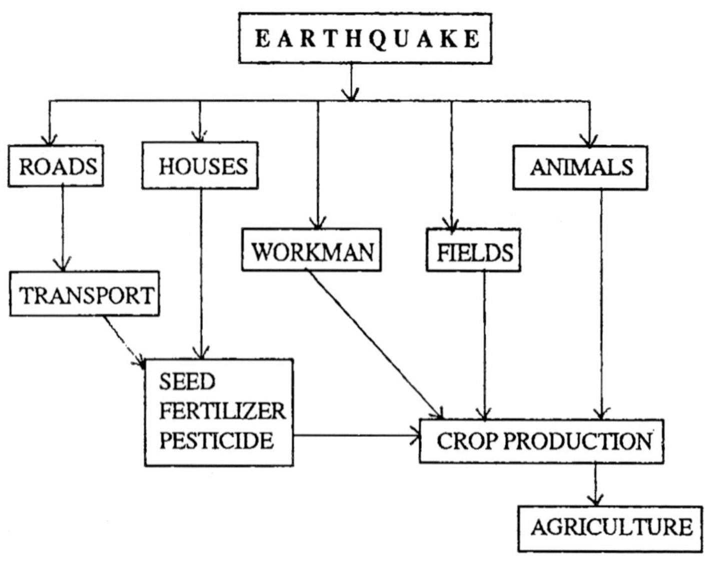

Fig. 6.3.1 Impact of earthquake on agriculture

6.4 IMPACT OF EARTHQUAKE ON TOURISM

Tourism is an activity of hills and playing a vital role in the economics and social development of these regions. It has been recognised as an agent for all developments. All the hopes of hilly peoples are linked with the development of tourism. Therefore it has been developed as an industry. Now question arises how an earthquake could affect the tourism? For giving an answer, let us take a look on the basic needs required for tourisms. These are proper approach routes, travel risk, accommodation facility, food supply and proper sanitation.

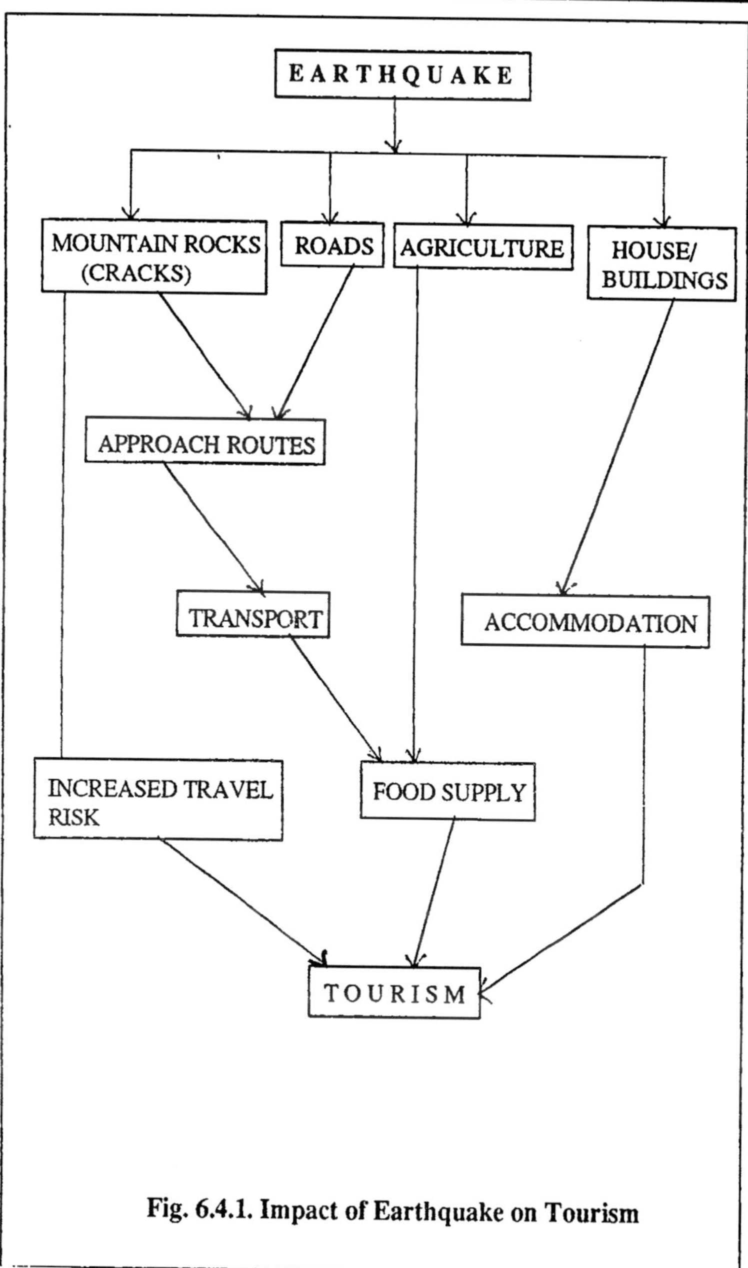

Fig. 6.4.1. Impact of Earthquake on Tourism

PROPER APPROACH

Today every one requires transport to travel and transport requires proper approach. In the hills transport facility is quite good but an earthquake may affect them badly. It was experienced in Uttarkashi when an earthquake struck this region on 20th October 1991. Out of 816 km, 72 km. road was ruined out. This damage could affect the transport and hence in turn restrict the number of tourists.

TRAVEL RISK

Travel risk in hills is always high. Every tourist wants to travel with the minimum travel risk. An earthquake could increase the probability of risk due to alarming cracks in rock structures, appeared because of the earthquake shocks. Therefore, fear of travel risk could affect the number of tourists adversely.

ACCOMMODATION FACILITY

Accommodation at tourist places is a basic need of every visitor. Already there is a shortage of accommodation to fulfil the requirement of tourists. If the house/buildings get collapsed, the situation could be aggravated further.

FOOD

Besides the accommodation food is also a necessary basic requirement of every person. If food supply get affected by any means it could result in the reduced tourism activity. After an earthquake food supply get affected by two means in hills, firstly due to low yield of the region and secondly due to lack of transport caused by the affected approach routes. Second mean may remained either for a short time, or for a long time depending upon the extent of destruction.

6.5 IMPACT OF EARTHQUAKE ON THE ECONOMIC RESOURCES

Earning of money in any region is basically based on various economic resources like agricultural production, industrial production, small scale industries, tourism etc. For understanding the effect of earthquake on the economic resources let us take an example of hills. In the hills various economic resources are horticulture, packing case industry, wool industry, handicraft industry, natural medicine, tourism and employment. Probable impacts of earthquake on these economic resources are being discussed in this chapter.

HORTICULTURE

With the development of tourism industry the horticulture is developing. The land of hills is most suitable for the development of horticulture and has great potential for producing vegetables and fruits. An earthquake may affect horticulture by various means, these are shown in Fig. 6.5.1.

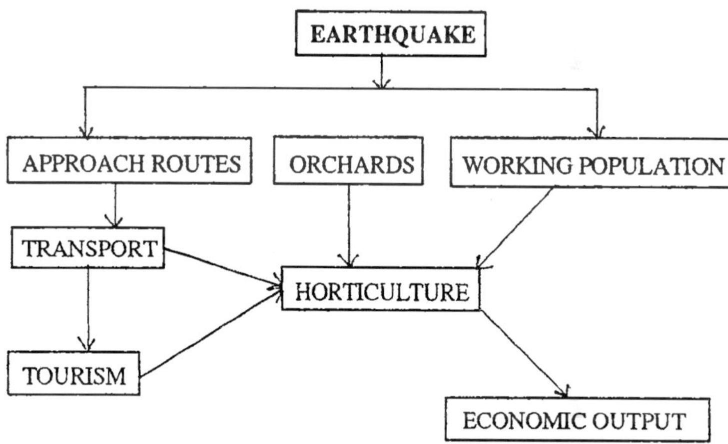

Fig. 6.5.1 Impact of Earthquake on the Economic output from horticulture.

PACKING CASE INDUSTRY

With development of horticulture, packing case industry has also come up. If horticulture get affected certainly it will affect the packing case industry (Fig. 6.5.2).

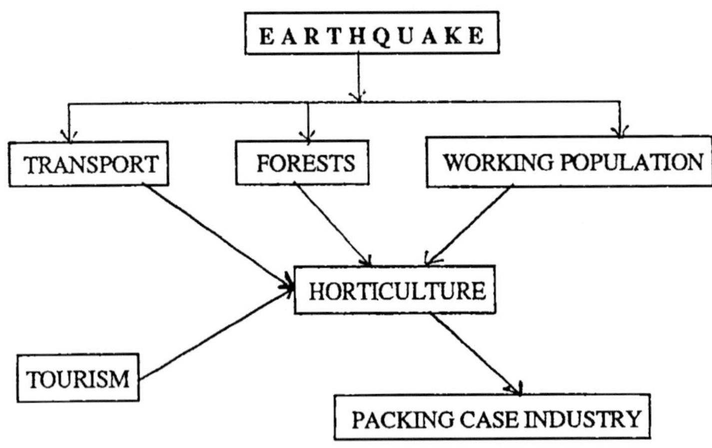

Fig. 6.5.2 Impact of earthquake on packing case industry

WOOLLEN INDUSTRY

Hilly regions are famous for their woollen industry where all type of woollen clothes i.e. carpets, shawls, blankets, sweaters etc. are being produced. The woollen industry in hills could get affected adversely from the earthquake disaster. Various factors which could affect this industry adversely are shown in Fig. 6.5.3.

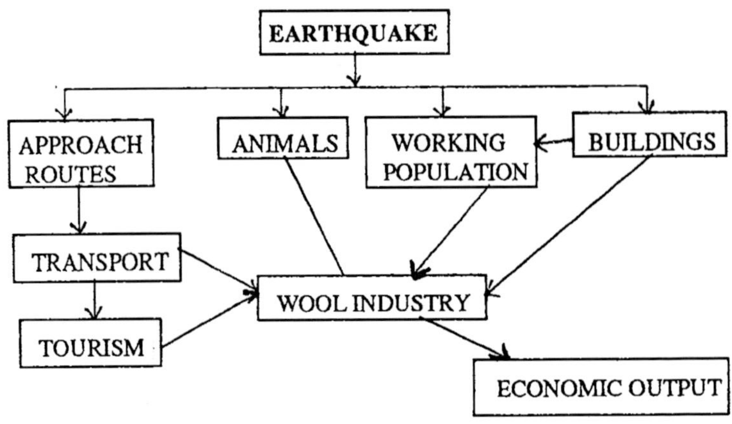

Fig. 6.5.3 Impact of earthquake on wool industry

HANDICRAFT INDUSTRY

It also get affected by the earthquake. Basically, the units are based on raw-material from forests. Besides the market availability of these products tourism also creates a new dimension for this industry in hills. Earthquake could affect this industry in various ways like transport, forest and market. (Fig. 6.5.4)

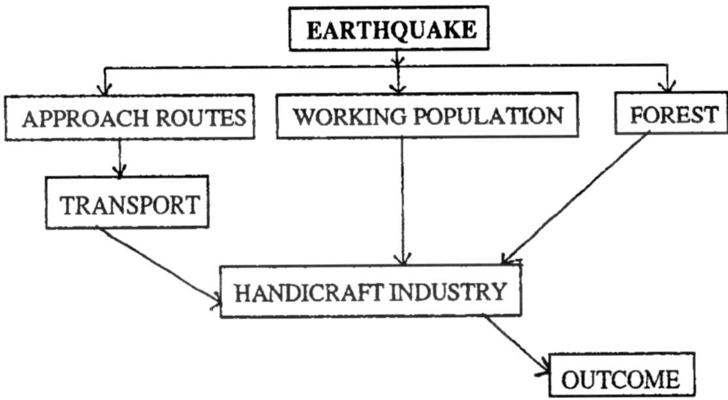

Fig. 6.5.4 Impact of Earthquake on Handicraft

NATURAL MEDICINES

The hills offer a great potentialities in this particular field of activity. Some natural medicinal plants are Barberries, Discorea, Deltoidia, Costus, Jatmansi, Serpentina, Timru, Vajradanti etc. The selling of these medicinal plants contributing to the economy of hills. But due to ill road network and transport, it could result in low economic output, as most of the natural occurring medicines are being obtained via difficult terrains. Shilajeet is found on the high altitude hill tops. To take out or carrying of these is a difficult task and after an earthquake it could become a very difficult due to various factors such as damaged approach routes, affected population etc.

TOURISM

In previous chapter the impact of earthquake on tourism has been explained. In the present day context tourism is contributing a major portion to the economic out put of hills. Various industries like hotel, transport and employment are based on the seasonal traffic of tourists, and contributing to the economic output of hills. Hence it is clear that ill tourism could result in low economic yield from tourism based industries.

EMPLOYMENT

The third major direct economic effect of ill tourism which is to be very vital disadvantage to hills relates to employment. The tourism is a labour intensive service industry. Therefore, it is a valuable source of employment. It employs a large number of people and provides a wide range of jobs which extend from the unskilled to the highly specialized. Tourism is also responsible for creating employment outside the industry. Hence, it is clear from the above statement that ill tourism, a result of earthquake, will certainly reduce the probability of high employment, which could result in the migration of local people towards other parts in search of employment.

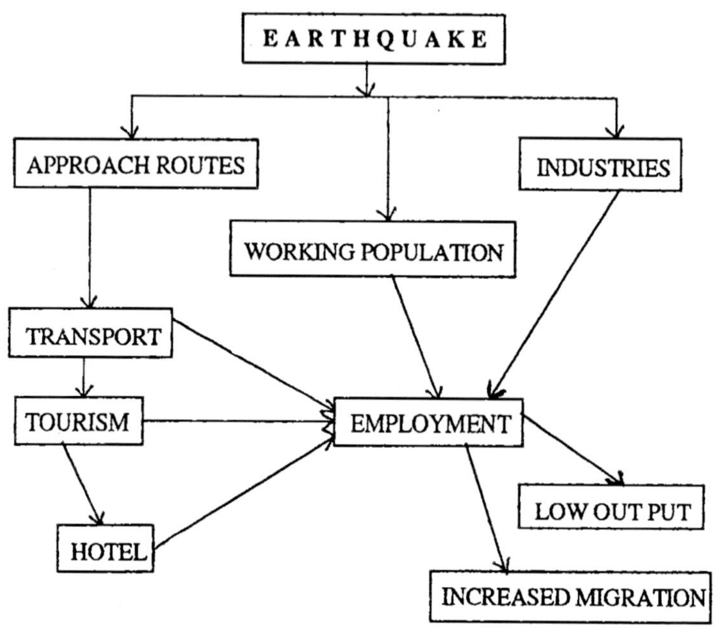

Fig. 6.5.5 Impact of earthquake on employment.

7
Houses and Earthquake

Earthquake is a natural disaster and cannot be stopped by any means. Therefore it is necessary to have a knowledge, why earthquake become so disastrous like that of 20th October 1991 and 30th September, 1993. As had experienced in the past major cause of great destruction is improper construction of houses and buildings without caring earthquake resistency. Major reasons which could aggravate the damage to buildings are being discussed hereunder.

IMPROPER CONSTRUCTION

Due to inadequate knowledge of the appearance of earthquake major portion of Indian population which is living in villages use indigenous methods to construct their houses. These methods does not have any proper advanced technique. Such type of constructions are unable to withstand earthquake shocks and come on the ground.

IMPROPER BUILDING MATERIAL

In most of the villages construction material used for houses is quite improper. Peoples generally used locally available stones or mud bricks as building unit and mud or lime as binding and plastering material. Such type of construction is quite prone to earthquake shocks. The houses of Garhwal and Maharashtra were also made up of the same material therefore, these get damaged easily.

DOUBLE STORY CONSTRUCTION

Walls of double story construction have an extra pressure, therefore, if construction material is improper walls are unable to withstand earthquake shocks and collapse more easily. It has happened during the earthquake of Garhwal in 1991.

AREA OF ROOMS

Poorly constructed houses having larger rooms are also more sensitive to earthquake shocks and have greater chance of collapse. Alongwith this, number of windows and doors also decreases the strength of walls.

ROOFS

If roofs are not constructed properly then it is another call for destruction. Putting of angled rafters on the walls exert an extra pressure on the walls of house due to which walls are unable to take care of earthquake shocks. It was observed in Garhwal during the appearance of 20th Oct. 1991 Earthquake.

FOUNDATIONS

If proper care for adequate foundation is not been taken during the construction of any building, it may increase the chance of collapse manyfold. As had happened in Garhwal on 20th October 1991.

PROTECTION MEASURES

To protect ourselves from such type of destruction it is necessary to construct earthquake resistant structures using appropriate construction material. For which we must take care of few following points.

— We always use 1 : 6 cement-sand ratio in place of mud and lime

— We keep appropriate foundation

— Use always either cement bricks or cooked bricks in place of stones, gravels and mud bricks.

— Always keep minimum windows and doors.

— Roof structure must be earthquake resistant, do not keep slate stone in tilted position on roof supporting structures.

— Keep the area of rooms smaller.

— Or construct temporary wooden houses in earthquake prone regions.

8

Prediction of Earthquakes

Whenever an earthquake struck any region one question arises immediately, can earthquake be predicted ? A few years back earthquake prediction was not even discussed seriously by seismologists. Due to the occurrence of two major earthquakes at Nagara and Alaska between 1964 and 1965, society forced scientists under pressure to study and develop an appropriate method for predicting earthquakes. Various theories and methods are existing but there is still no perfect method we have for predicting earthquakes.

Prediction of earthquake may be categorised into three major groups i.e. prediction of earthquake, 10-20 years before (long-term prediction), prediction of earthquake before 2-3 months (mid-term prediction) and prediction 3-4 weeks before (short-term prediction).

Earthquakes may be predicted on the basis of following observations.

- Change in the earth surface and earth crust
- Change in the speed of earthquake waves
- Change in the geomagnetic field
- Change in the electrical field
- Change in the level of ground water
- Change in the river flow, river courses, and water quantity
- Change in the frequency of earthquakes
- Anomalous behaviour of animals.

According to the Russian scientists 'P' and 'S' waves can be differentiated even after travelling a long distance from the focus. The ratio between the speed of these two waves get disturbed before the appearance of earthquake. Hence, by observing such changes an earthquake could be predicted. According to few scientists earthquake could be predicated by analysing ground water. Sudden increase in salt and mineral contents in ground water is an indication of earthquake. Presence of high concentration of argon in ground water is also an indication of earthquake. It could also be observed 2-8 days before. According to another assumption earthquakes could also be predicted by analysing gaseous emanation. High ratio between argon and helium and argon and nitrogen is an indication of the occurrence of an earthquake.

Another method which has been developed is seismic gap method, based on the principle of earthquake repeatability i.e. after a definite time interval is a particular region, earthquake repeat its occurrence. One more method is seismic grid method, it is based on the observation of number of shocks per grid per magnitude per decade, which is supplemented by physical and mathematical models and satellite pictures.

GEOMAGNETIC FORCE AND EARTHQUAKE PREDICTION

According to the scientists, temperature of earth surface (rocks) surrounding epicentre get increased due to the force, stress and the movement of rocks below the earth surface. Due to this increase in rock's temperature the geomagnetic force of the nearby area decreases which in turn affects the presence and concentration of positive and negative particles in the environment. Radio waves always travel with the help of these particles, therefore, radio transmission could also get affected in the area surrounding epicentre. This effect could be observed quite before the time i.e. 10-30 hrs before. The indications are like

- Radio will not catch transmission of a particular station on its daily frequency, suppose radio catch a station 'A' on its normal frequency of 1000 khrtz every day but before the earthquake it may appear between 1100 and 1500 khrtz.
- A continuous sound disturbance on the radio.
- Same as above a continuous disturbance in picture and sound on T.V. Sets.

— Other electronic instruments like telephonic and telegraphic also produce abnormal characteristics.

ANIMALS AND EARTHQUAKE PREDICTION

On the one side seismologists are still engaged in developing an appropriate method to predict an earth accurately with regards to place, time and magnitude. Animal behaviourists, on the other side are also trying hard to determined whether the anomalous behaviour of animals could be used to predict an earthquake.

According to the past experiences some animals are much sensitive than humans to perceive kinds of geophyical stimuli which may occur before earthquakes. Such type of animals react within the epicentral areas. The time of indication may vary from 10 to 30 hours.

Most common animals which comes under this category are dogs, horses, cows, and chicks. Other animals also react but only few hours before of the earthquakes. Fishes and rats are also quite sensitive to earthquakes. Fishes are reported to have responded to quake of 2.0 magnitude on Richter scale. Chimpansees and other primates are quite sensitive to gaseous emanations preceding earthquake, these can sense such emanations even in extensive dilutions. Every change before the earthquakes affect animals in their own ways. During two different earthquakes in a particular region and in two different regions same species of animals could behave in similar manner, just because of the difference in precursors released by different earthquakes.

There are various examples of anomalous behaviour of animals, experienced in the past, are known. In the February 1975 earthquake in China rats were agitated and were easily caught, snakes come out of hibranation, small pigs bite each other, cows fought and dug the ground, deer run away, turtles jumped out of water and cried. In 1905 Kangra of Himachal Pradesh animals of Zoo situated 180 kms. away from the epicentre (Lahore) started crying in midnight before the earthquake. Also before October 20th 1991 in Garhwal and surrounding areas cows, dogs and cats were found upset and cried whole the night.

Other than above stated precursory signs whenever there is an earthquake comes few more characteristics may also be observed

— Walls of houses tilt slightly

— Small rivers change their course
— Rail tracks get twisted and bent
— Water pipes became defective
— The axis of machines in an industry tilts

9

Quake Predictions Get a Jolt

— Rajendra Prabhu

Earthquake study and prediction may be hard science; but astrologers and geophysicists have one thing in common — for each success, there are as many instances when they often miss their target. That, in a way, explains why two eminent scientists in geophysics were contradicting each other soon after the big tragedy at Latur. Here are some examples of predictions gone astray.

Buried somewhere in the annual report of the prestigious National Geophysical Research Institute for 1990-91 is this sentence.

"Preliminary assessment of the location of June 1990 shocks indicates their proximity to this area only. Confinement of the epicentres in the same area is an encouraging fact. If the recent activity had shifted it would indicate the possibility of a strong earthquake."

The reference is to the study of reservoir induced seismicity in Bhatsa dam area. It is 80 km from Bombay. The actual happening was 350 km away at Latur. Nor was it a reservoir induced earthquake. Scanning through the annual reports of the NGRI for the last three years there is little evidence that the possibility of pressures building up along the Deccan fault in central Maharashtra area was at all within the ken of the NGRI investigations. There seems to be some truth in the allegation made by former NGRI Director Dr. Vinod K. Gaur that "there has been no investigation which can tell us about the location and seismic activity of the (Deccan) fault and its disposition".

As against this the Himalayan region has been under intense observation, particularly North-east which is considered the most vulnerable part of the country to earthquakes. Yet, not only the Uttarkashi earthquake of 1991 October was not foretold, when it actually occurred the Indian Meteorological Department mislocated the epicentre by as much as 100 km from actual location. "It is so distressing" says Dr. Gaur.

The seismic study of this country is dotted with a multiplicity of authorities and a multiplicity of instruments. The Department of Science and Technology funds many programmes including an intensive study of the Himalayan regions. The IMD keeps a tab on the tectonic movements through over 50 facilities spread over the country. The Department of Atomic Energy has its own stations at Gauribidanoor (Karnataka) and certain other places in strategic locations mainly meant to tab any nuclear explosions taking place in neighbourhood. The NGRI is doing seismic studies mainly in Himalayan frontal areas including North-east and also in reservoirs like that in Koyna, Nagarjundsagar etc. It also undertakes seismic studies for the ONGC and the Department of Atomic Energy. There are still other authorities including military which have their fingers in the seismic pie.

Varying Reliability: All these various authorities have their own instrumentation with varying degrees of reliability and caliberation. The NGRI has obtained the latest digital telemetry for its scan of the North-east and got imported state-of-art equipment for the seismic survey of Koodamkulam in Tamil Nadu where a nuclear power plant was proposed to be located. The IMD's instruments in different locations are of much older generation. There is no recaliberation of all these various instruments so that a single benchmark scale could be evolved for the entire country. Some of the imported equipment are caliberated as per the American practice which is quite different from the British standards that we normally follow.

The philosophy that seems to guide the policy making is that managing disasters is better than preventing them. Of course, at the present level of geophysical science, earthquakes cannot be predicted with any certainty. But its directions can be measured in advance and precautions taken. For instance the same NGRI faulted for inability to get even a hang of the developing situation in the Deccan trap, has done excellent work in the Himalayan and Indo-Gangetic region.

Equipment like three strong motion seismic arrays in Himachal, UP and North-east have given excellent data with which Dr. Harsh K. Gupta has forecast that major earthquakes in the area will continue to occur. With more data and refinement there is every possibility of determining the critical sites in the Indo-Gangetic belt and thus reduce damage if adequate steps are taken before. Similar studies could be undertaken for the areas on peninsular India, also given the sophistication in the instrumentation available if money could be found.

Earthquakes in India have a different cause than what geophysicists abroad are familiar with. To illustrate what is happening geophysicists suggest this simple experiment. Take two stone plates and join them edge to edge from the north and south. Apply pressure on the northern and southern edges. There would be intense cracking along the two edges which press against each other. Imagine the stoneplate at the south as the Indian plate (or Gondwana plate) which drifted out from southern Indian Ocean many millions years ago and is being pressed against the Tibetan (Eurasian) plate. The Himalayas, the youngest mountains in the world arose out of this collision virtually from the sea and they are still rising as the pressure increases. Naturally the Himalayan region has been the major focus of all seismic studies in the country, specially by the NGRI. The northward movement of Indian plate is at the rate of 5-1/2 cms per year.

Plate Anology: If we take the stone plate anology a little further greater pressure on these colliding plates from opposite directions would see many cracks developing along them. There have been many cracks along the western edge of the Indian plate or what are called Deccan faults. The Latur disaster is the clinching evidence that the Indian plate is developing pressures along the central portion. The entire Godavari basin running from western Maharashtra to eastern Andhra Pradesh is likely to be affected. The Latur incident confirms, say geophysicists, that henceforward the attention of scientists should also be concentrated along the central portion of peninsular India. NGRI has already moved to extend the programme of the morphostructural zoning map of the Himalayas to entire country.

Funds for such research are difficult to obtain. The NGRI's entire budget spanning many disciples besides earthquake hazard measurement, is just Rs. 8 crores. The reluctance to allot funds for such work is seen in the long battle for the acquiring of the Geodetic very long Baseline Interferometry equipment.

The NGRI has already established through experiments the need and usefulness of this equipment in measuring acceleration of the tectonic plate at various points in the country. A correct value of acceleration is necessary for calculating the earthquake hazards for, among other things, dams of heights more than 100 metres at specific points. All over the country there are over 100 such sites and crores of people are exposed of dreadful scenarios like the Koyna dam disaster. Yet, because some Rs. 15 crore are not available the whole programme is stuck. The same Government which fails to sanction Rs. 15 crore for an equipment to help estimate the hazard would sanction hundreds of crore of rupees once the disaster strikes as has happened at Latur now.

In such cases, the bureaucracy has a stock answer: no funds. And if one presses if further, they ask: what guarantee is there that VLBI would predict earthquakes? Certainly VLBI would not, but it would help estimate the hazard accurately and thereby give adequate warning. For instance, NGRI has already established the relation between seismic activity and the loading of reservoirs. Out of such studies at various dam site, the guidelines for safe filling the reservoirs have been evolved.

Mapping and monitoring of all parameters that cause earthquakes seem to be the only alternative till science gives us a different means of accurate and reliable prediction of earthquakes. The actual job of predictions could be vitiated by the fact that sometimes a period of quiescence precedes the storm. But the state of science today can locate critical areas, can keep all the parameters closely watched and give a fair indication of what could happen if earthquake comes.

A probabilistic seismic hazard map of Himalaya and adjoining regions was prepared recently by Mr. K.N. Khatti for a period 1983 to 2031 which brings areas like Delhi within the peak acceleration of 20 per cent of "g" (32 ft per second) which is serious enough. Most of the North-east is beyond 70 per cent of "g" acceleration almost the same as on the crest of the Himalayas that is, highly hazardous. The danger zones have already been marked with some degree of reliability.

Dr. Harsh Gupta and Dr. K.N. Singh predicted in 1986 the possibility of a major earthquake of magnitude 8.5 in the north-east based on study of various data. This came true in 1988. These instances are encouraging and point towards what could be done to minimise the dangers from these natural disasters.

Liquification potential : Soil liquification potential during earthquakes is another study that might help to minimise the dangers from earthquakes. A beginning has already been made by a DST programme to study soil liquification potential of selected areas including Delhi which has reported high potential. Such studies help take preventive action like building codes being formulated to prevent concrete structures from collapsing.

In fact, it has been found that if building codes follows measuring of data regarding soil structure and earthquake damage potential of the area, thousands of human lives could be saved. It has happened in Japan. The Uttarkashi earthquake showed that buildings which were solid structure built without being made earthquake proof, were responsible for many deaths while many poor victims escaped death due to their houses being light. In the cyclone prone areas of Andhra scientific studies proved that simple constructions build to withstand storms helped save lives.

How important such preventive steps are, can be seen from the juxtaposition of data from the Kangra region where an earthquake occurred in 1905. An estimate made by A. S. Arya in Current Science issue says that if similar earthquake were to occur now, depending upon the time of the occurrence, the death toll would vary between 88,000 and 3,44,000. With population density growing and more houses being of stone and concrete than before the danger potential is increasing every year all along Indo-Gangetic belt.

The studies of earthquakes cycles along what is known as the Alpide belt stretching all the way from Greece to Indonesia through the land route of Turkey. Iran, Baluchistan, Indo-Gangetic belt etc. are ominous. After a relative quietude of 1952-1991, there is every possibility of seismic activity being more intense in the coming thirty years. The greater reason for the alarm bells to ring.

Courtesy : The Hindustan Times

10

What we can do ?

We have faced so many devastating earthquakes in the past and knows the fact that it is a natural phenomenon which could not be stopped, it is necessary to take few preventive measures to stop further destruction and causalties during such event. For this purpose few following actions must be taken on immediate basis.

- — Reidentification of earthquake prone regions
- — Government must inforced building regulations to construct earthquake resistant buildings in earthquake prone regions.
- — All the peoples residing in earthquake prone regions must be informed and trained.
- — Reexamination of dam's safety which are under construction in earthquake sensitive regions must be done.
- — All the population living in the earthquake must be trained to take protection measures during earthquake.
- — An appropriate method to predict an earthquake accurately i.e. date, time and location must be developed.

PROTECTION MEASURES DURING EARTHQUAKE

- — On the receiving of first shock come out from building and stay in open place.
- — If unable to come out sit below the table or any other hard structure.
- — If you are driving a car stay in car.
- — Do not run towards buildings.
- — Do not ignite any flammable material shut off all the flames immediately.
- — If you are having helmet wear helmet on head.